OAKWOOD LIBRARY OF RAILWAY HISTORY

THE
FURNESS RAILWAY

R. W. RUSH

Furness Railway Staff. *Oakwood Press*

THE OAKWOOD PRESS

© R. W. Rush, 1973

ISBN 978-0-85361-773-0

First edition 1973.

Printed by
Claro Print, Office 26, 27, 1 Spiersbridge Way
Thornliebank, Glasgow G46 8NG

This second edition 2024, combines the original text from R. W. Rush's OL 35 *The Furness Railway* and OL 35A *The Furness Railway: Locomotives and Rolling stock.*

Robert Rush's engine classes are controversial and as he writes in this book:

"There was no official classification of locomotives. There is, however, a classification in existence which has been more or less accepted, is believed to have been introduced by that eminent historian, the late A. C. W. Lowe. As this classification is very useful in sorting out the different types of locomotives from the very early days, it will be continued here, but it must be emphasised that it is not official."

The classification remains more or less accepted but several of the classes Rush describes can be broken down further. The D5 and K4 classes are examples. There are hints in his text that each class could be subdivided but he doesn't elaborate. Similarly the Furness Railway didn't use diagram numbers for rolling stock and the ones Rush uses are from a later (1930s) post-Grouping book used by the LMS to categorise the stock it inherited from the Furness Railway.

OL 35A contained a table of locomotive dimensions and another two of Furness Railway rolling stock diagram numbers. Because more recent work has largely superseded these, they are not reproduced in this edition. The Cumbrian Railway Asscociation has compiled lists of these details, from new research, and these are available to their members through their website.

In the fifty years since this book was first published there have been few others on the Furness Railway. Of this small number *The Furness Railway – a history* by Michael Andrews stands out. It describes the railway in significantly more detail than this short volume can, but is now out of print. However, it is worth finding a second-hand copy of if this book has whetted your appetite.

The maps on pages 8 and 13 are reproduced with the permission of the National Library of Scotland. The images are licensed under the Creative Commons (CC-BY) licence. to view the licence visit https://creativecommons.org/licenses/by/4.0/

Published by
The Oakwood Press, 54-58 Mill Square, Catrine, KA5 6RD
Telephone: 01290 551122, Website: www.stenlake.co.uk

Contents

Preface		4
One	The Town and Port of Barrow-in-Furness	9
Two	The Furness Railway	21
Three	The Whitehaven & Furness Junction Railway	49
Four	The Whitehaven, Cleator & Egremont Railway	57
Five	The Furness Railway from 1880	63
Six	The War Years and the Grouping	79
Seven	The Cleator & Workington Railway	89
Eight	Furness Railway Locomotives	99
Nine	Whitehaven & Furness Junction Railway Locomotives	145
Ten	Rolling Stock	149
Eleven	Accidents	165

Appendices

One	Locomotive List	173
Two	Gradients	182
Index		183

A group of Edwardian holidaymakers at Foxfield. *Oakwood Press*

Preface

This book has been by no means an easy proposition; in fact it has twice been completely rewritten, and in all has taken rather more than eleven years to produce. When delving into railway history, one finds that surprisingly little has been written down concerning the Furness Railway, and what has been written has proved to be largely repetitive. The author despaired of finding anything new on the subject until, quite by chance, he came upon a book in the local Public Library which dealt with the social and industrial history of the district. In this, *Furness and the Industrial Revolution,* by Dr. J. D. Marshall, were found numerous references to the Furness Railway which proved invaluable in writing this present book, although these references were merely incidental to the main theme of Dr. Marshall's treatise. In addition, other works which have been consulted, including the late McGowan Gradon's *Furness Railway,* and standard publications such as *The Locomotive,* the *Railway Magazine, Railway World, The Model Railway Constructor,* and *The Model Railway News,* have all provided various references and data which have proved invaluable.

Also, the author is greatly indebted to a number of gentlemen who have helped considerably in their various ways towards the compilation. In particular, Mr Wyndham Lee, who drew on his personal memories of the line, and who, incidentally, carried out the first rewrite of the book himself. He was also kind enough to allow me to draw on his vast collection of photographs. Secondly, Mr A. F. N. Barnsdale, who has contributed in no small measure to the locomotive history from his extensive records of the motive power of the LMS. Also to my cousin, James Ranson, who carried out a great deal of research in the Barrow-in-Furness Public Library on my behalf. Mention must also be made of Mr R. M. Casserley, who has helped very greatly in attempting to unravel the very vexed question of the Furness Railway passenger stock. Without the cheerful collaboration of these four gentlemen, this book would never have been possible. In fact, at one period the author completely gave up the project in despair, and it lay dormant for almost two years. After considerable urgings from his collaborators, he consented to carry on, and so was born the second complete rewrite. Even so, it is admittedly not completely satisfactory in the author's eyes, but it seems impossible to make any further improvements.

Some details of the locomotive history, mainly in regard to the Whitehaven & Furness Junction and Whitehaven, Cleator & Egremont Railways, have been a real bone of contention, and may not therefore be entirely authentic, but confirmation has not been available. The version of the W&FJR locomotive list compiled by the late A. C. W. Lowe was

PREFACE

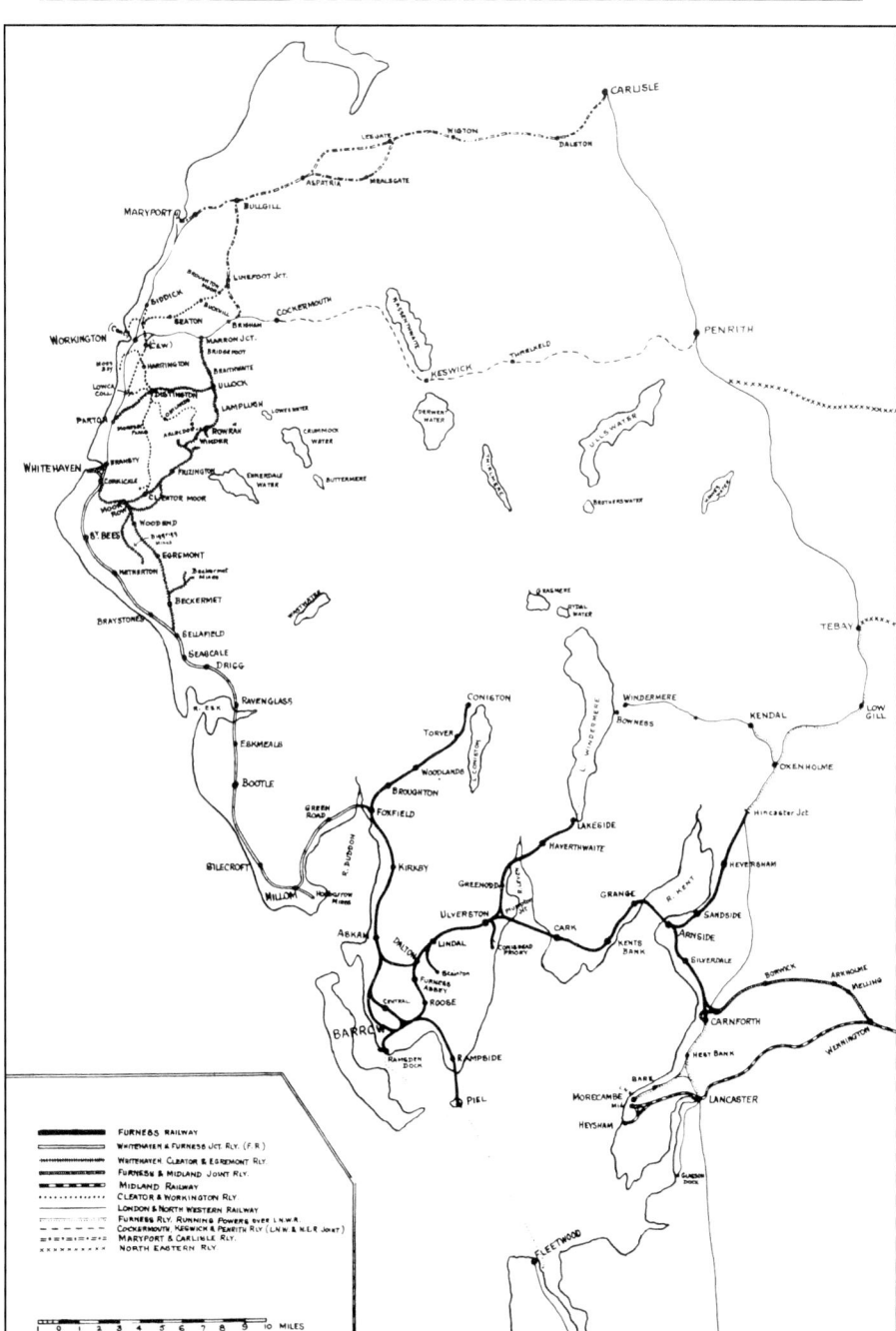

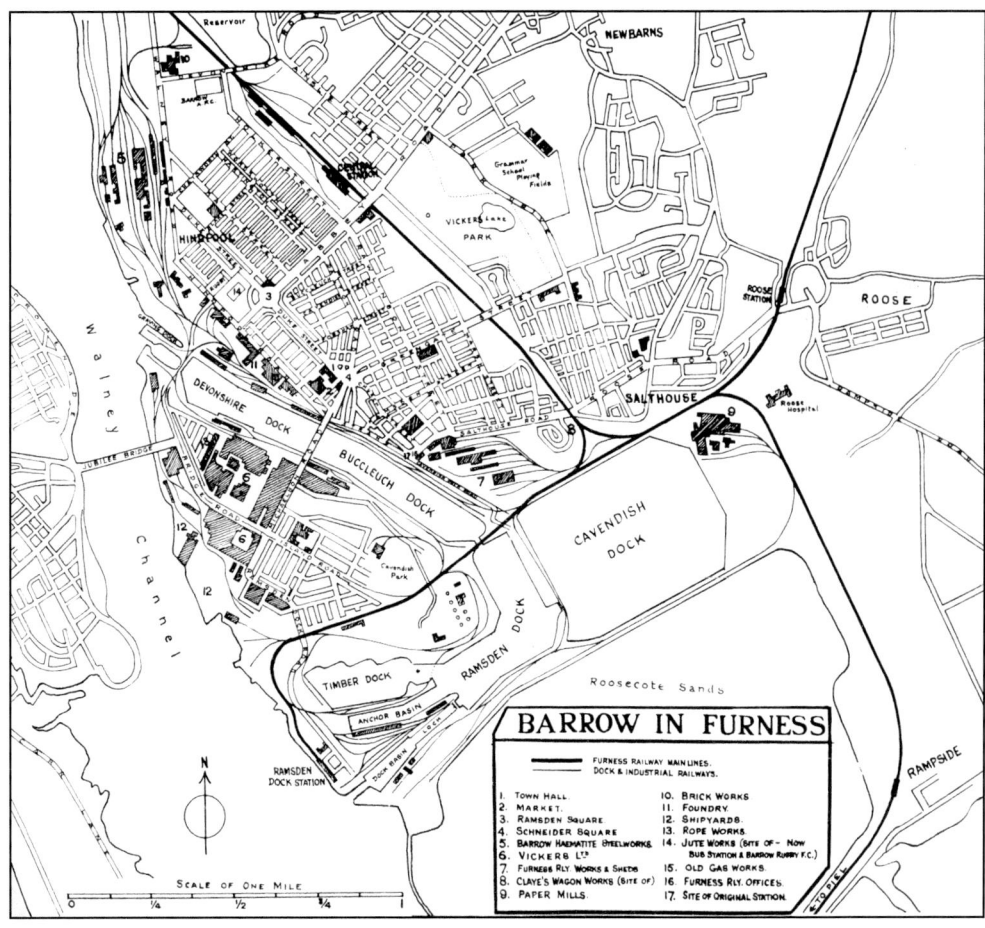

generally accepted and used as a basis for the list given, but this had to be considerably modified by facts which have come to light through Mr Craven's researches into the early locomotives of the Lancashire & Yorkshire Railway, published in the journal of the Stephenson Locomotive Society, from which it has been established that some old L&YR engines were sold to the W&FJR.

The passenger rolling stock, as mentioned earlier, has been an enormous problem, which, at this stage, seems incapable of solution. Due to the loss of a large part of LMS records by fire in 1951, no complete list of Furness Railway stock now exists, and though a certain amount of LMS renumbering, both in 1923 and 1933, has been possible to correlate, the complete list cannot be reconstructed. Only odd Furness numbers are known, but it is almost impossible to relate these to their corresponding LMS numbers. The drawings in the supplement have been made from the official Furness diagrams, which, it must be said, were in many cases rather crude, and the opportunity has been taken in the line drawings to tidy them up. Even so, some minor details are the subject of some surmise. Dates of building and totals have also been taken from these same diagrams. Earlier details, to 1900, of the locomotives have been obtained from W. F. Pettigrew's paper to the Institution of Locomotive Engineers.

In conclusion, it is hoped that this book will clear up some of the earlier errors which have, unfortunately, been perpetuated in print from time to time and have come to be accepted as fact.

Robert W. Rush, Accrington, Lancs.

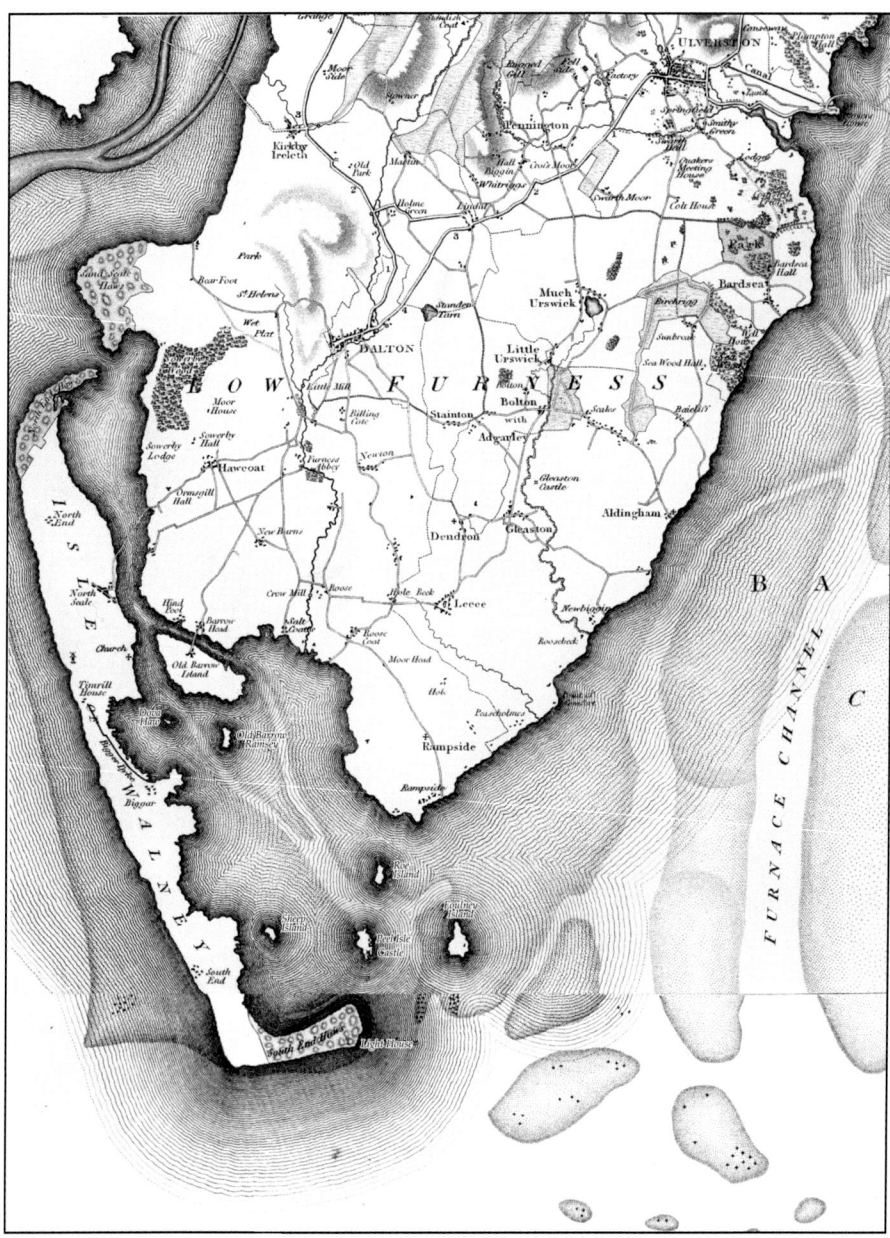

Barrow and Low Furness in 1818, from *Map of the County Palatine of Lancaster from an actual survey* by Christopher Greenwood of Wakefield.
Reproduced with the permission of the National Library of Scotland.

Chapter One

The Town and Port of Barrow-in-Furness

On the far shores of Morecambe Bay, nestling under the Cumberland Fells, is that remote corner of England known to the Vikings as "the further ness", from its being so isolated from the main centres where they settled. This corner is separated from the main County Palatine of Lancaster, to which it belongs, by an isthmus known as the Hundred of Lonsdale North of the Sands, and which contains such centres as Cartmel and Grange-over-Sands, and also by the Rivers Kent and Leven, the former flowing down from the Westmorland Fells, the latter forming Lake Windermere's outlet to the sea.

After the demise of Furness Abbey in the 16th century, Ulverston, with a population fluctuating between 800 and 1,100 persons, formed the principal town and market of Furness. Barrow, at this time, was a small hamlet of fisherfolk, settled on the shore of a sound between Barrow Island and the much larger island of Walney, at a point where the occasional ferry would cross the channel. Its population was only about 150. The nearest settlements otherwise were at Rampside, where there was a deep water roadstead between the mainland and the twin islands of Piel and Roa, and at Dalton, some four miles distant through the grounds of Furness Abbey.

A postcard produced by the Furness Railway of Furness Abbey. Their caption on the address side reads:- *The magnificent ruins of Furness Abbey are situated at Barrow-in-Furness. Visitors should book to Furness Abbey Station. Furness Abbey Hotel is beautifully situated within the Abbey grounds.* *Oakwood Press*

During the 18th century, the development of the horse carriage as a normal means of transport led to the development also of the coaching services from the Bear & Staff Inn, Lancaster (whence there was a connection from the south) and the Sun Inn, Ulverston, a journey performed over the sands in a mean time of nine hours, depending on weather and tide. Not until the opening of the Ulverston & Lancaster Railway in 1857 did this coaching service cease to operate.

The steam railway, as a means of communication, was first introduced into the district in 1846, although there had been various attempts previously to establish tram roads to serve the network of mines around Dalton. This original line had a total length of 18 miles, and was completely isolated from any other line, although it had been originally envisaged as part of a grand trunk route between England and Scotland, to be known as the Caledonian, West Cumberland & Furness Railways, first mooted between 1834 and 1838. At the turn of the decade, John Abel Smith, a partner in the Furness banking firm of Smith, Payne & Smith and MP for Chichester and Midhurst, purchased Roa Island, at the tip of Furness, after having heard of these various railway schemes, and being under the impression that he might profit thereby. At the same time, the Preston & Wyre Railway, wishing to further its services, proposed a steamship service from its terminus and harbour at Fleetwood to Roa

The 1¼ mille long Ulverston Canal connects from its sea lock on Morecambe Bay to the town basin at Ulverston. *Oakwood Press*

Island. In the year 1840, Smith introduced a Bill to Parliament in which he sought powers for the formation of a harbour company which would have wide control in the area, including the Walney Channel Roadstead, which at this time was still a creek of the Port of Lancaster. Opposition to Smith's Bill was raised by the Port of Lancaster Authority and by the Ulverston Canal Company, which had constructed a waterway from the shores of Leven at Ulverston Sandside into the centre of the town, where a basin afforded accommodation to vessels of large tonnage. However, it was not upon the strength of the canal company's opposition that the Bill foundered, but rather upon the insistence of the local shipping merchants, backed by the Lancaster Port Authority, that the plan would interfere with the ancient rights to Piel Roadsteads, rights granted and confirmed during the control of Furness Abbey over the district. Undaunted, Smith presented a second Bill to Parliament, by which he would be empowered to construct a causeway to the mainland at Rampside, officially "a causeway to link Roa Island with the Main Island of Great Britain". Further powers would be allotted for the improvement of the wharfage and pier facilities at Piel, and for the construction of warehouses. To further his interest, Smith purchased a small steamer and began a service between Piel and Fleetwood. The original plan of the Preston & Wyre Railway had in the meantime collapsed.

When, in 1844, the newly-created Furness Railway Company was preparing its proposal for lines in Lower Furness, Smith intimated that he would impose strict tolls for any trains which would pass over his causeway, should the Furness build a line along it. Because of this opposition, the railway company turned their thoughts to the hamlet of Barrow-in-Furness as a possible alternative terminus, for there was also good wharfage at their disposal. Smith was unable, however, to exploit his obstruction tactics much longer, for in December 1852, soon after the construction and opening of the railway, a violent storm badly damaged the pier and embankment at Piel; rather than face the great expense of repairs, Smith agreed to sell his whole undertaking to the Furness Railway for the sum of £15,000. The Piel trade did not always meet expectations, however.

Because of Smith's obstruction, the promoters of the Furness Railway had already decided on Barrow as their objective. Included in the original minutes of the company was a proposal, which was approved, for the improvement of the existing jetties and the laying of lines of railway upon them, together with such improvements as dredging and widening of Barrow Channel, to permit ships of larger tonnage to moor against the jetties. All these proposals were eventually carried out.

On 30th June, 1848, the Act of Incorporation of the Barrow Harbour Commissioners received the Royal Assent. By now the decline of the Port of Lancaster had weakened the opposition. The Barrow Act authorised the Commissioners to borrow £15,900 from the Furness Railway Company, and also stated that the Chairman of the latter's directorate should also be a Commissioner. During the ensuing five years, not much was achieved by the Commissioners, apart from spending a few hundred pounds on deepening Walney Channel. In June 1853, the Furness Railway's civil engineer, J. R. McClean, carried out a further survey, which allowed for the deepening of Barrow Channel by a further three feet, thus giving a depth of 18 feet at mean tide. This report was acted upon, but the necessary works were not completed until 1857, when at last vessels up to 500 tons could sail right up to the wharves.

A further Act of Parliament, the Barrow Harbour Act, 1863, gave the Commissioners even wider powers than those previously enjoyed, and permitted the borrowing of £137,000 for the construction of quays and docks in the Barrow and Walney Channels. James Ramsden, one of the motivating forces of the Furness Railway, and McClean together produced an ambitious plan for the construction of a stone wall, with lock gates, across both ends of the Barrow Channel, thus virtually joining Barrow Island to the mainland and forming the enclosed waters into a large dock. At its widest point the channel had a breadth of 300 yards. The contractors for the accepted scheme were Messrs. Brassey & Field, their tender being for £32,427. Almost four years passed before the scheme was completed, and much money was lost in the construction by the building firm. Theirs was the same problem as was met everywhere in Furness – shortage of labour. The dock, divided into two parts, was eventually opened in 1867, the two halves being named Buccleuch and Devonshire, after those Dukes who had played a large part in the development of the Furness Railway. The western shore of Barrow Island was set aside for shipbuilding sites, whilst the dock side of the island was to form a large timber yard, stretching the whole length of the dock.

The Barrow Shipbuilding Company was formed in 1869, and remained in operation until 1888; in the latter year, the company was incorporated into the newly formed Naval Construction & Armaments Company. A further eight years passed, and this company became the world-famous firm of Vickers, Sons & Maxim. Although many varied types of ships were built here prior to 1896, the yard became in that year one of the chief shipbuilding yards to the Admiralty, specialising later in larger classes of warships and submarines.

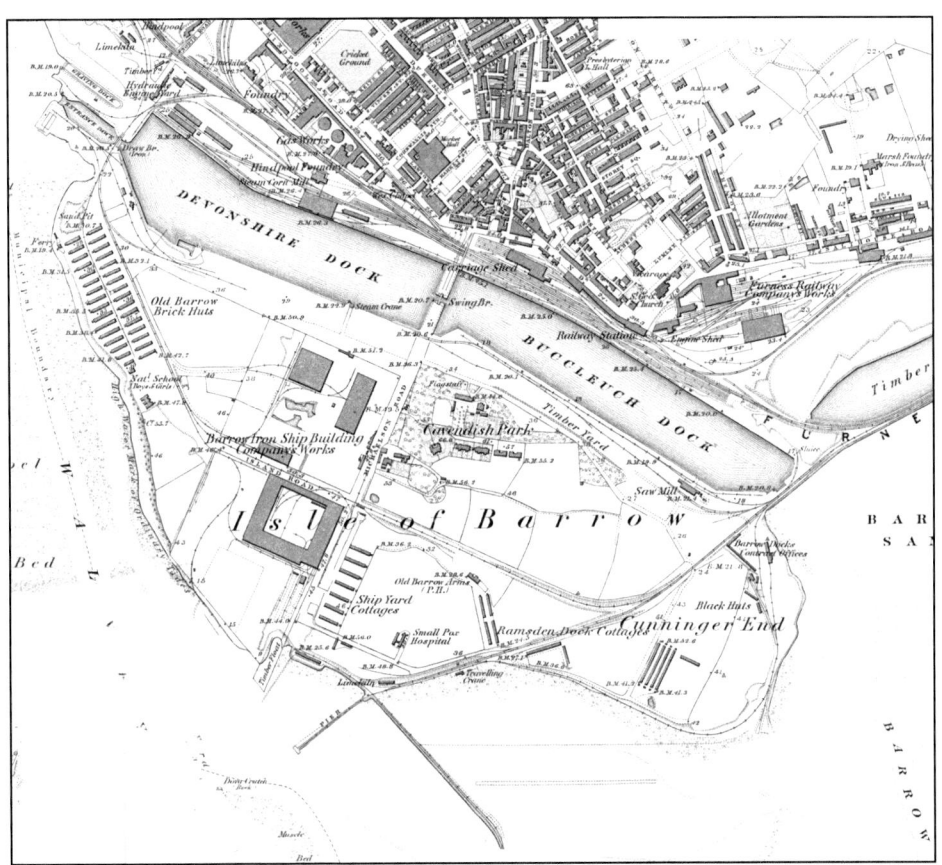

Barrow Docks in 1873.   © *National Library of Scotland, reduced from the 6 inch map, 2024.*

By enclosing a portion of Walney Channel at the south-western end of Barrow Island, two further docks were formed, Ramsden and Cavendish. The latter was to serve principally for the import of timber, whilst for Ramsden Dock it was proposed that passenger sailings be inaugurated. The half-yearly report of the Furness Railway directors for 9th August, 1880, contained the following notice: "The trade between Barrow and the Americas by the steamers of the Anchor Line is being satisfactorily developed, and the directors anticipate that a weekly service will shortly be commenced". In co-operation with the Midland Railway, there were also services to Belfast, Dublin, and the Isle of Man. Cavendish Dock was opened at the end of July 1878, while the sea was admitted to Ramsden Dock on 27th May, 1879. The total acreage of docks was now 234. The commencement of the construction of these two docks in 1875 helped to alleviate somewhat the period of depression of the mid-seventies, when the iron and steel trade was at a low ebb and many were unemployed. This depression, in various strengths, was to last nearly 20 years, and proved a very difficult time for both the railway and the town. The main exports from the docks consisted of pig iron, chiefly to South Wales, and steel rails destined for the construction of railways in the colonies. Imports were of timber from the Baltic ports, jute, Spanish iron-ore, and livestock, slaughterhouses and a "Foreign Animals Wharf" having been constructed as part of the layout of Ramsden Dock. The shipyards suffered acutely at this time, and as early as 1877 the Duke of Devonshire was persuaded to invest more capital to support them. Sir James Ramsden and Devonshire jointly invested more than a quarter of a million pounds in the yards during the late seventies. Strikes and labour disputes were rife at this period, the main causes being poor working conditions, poor accommodation, and bad pay. It was not until the turn of the 20th century that the shipyards and docks were able to escape from their difficulties.

Seen in the light of modern trends, the docks of Barrow were perhaps an unnecessary development; never during their existence have they been fully exploited, for they lie too far from the main centres of

HMS *Natal* being fitted out in Devonshire Dock in 1905. Alongside is HIJMS *Katori* also being finished in preparation for its voyage to Japan.   *Oakwood Press*

THE TOWN AND PORT OF BARROW-IN-FURNESS

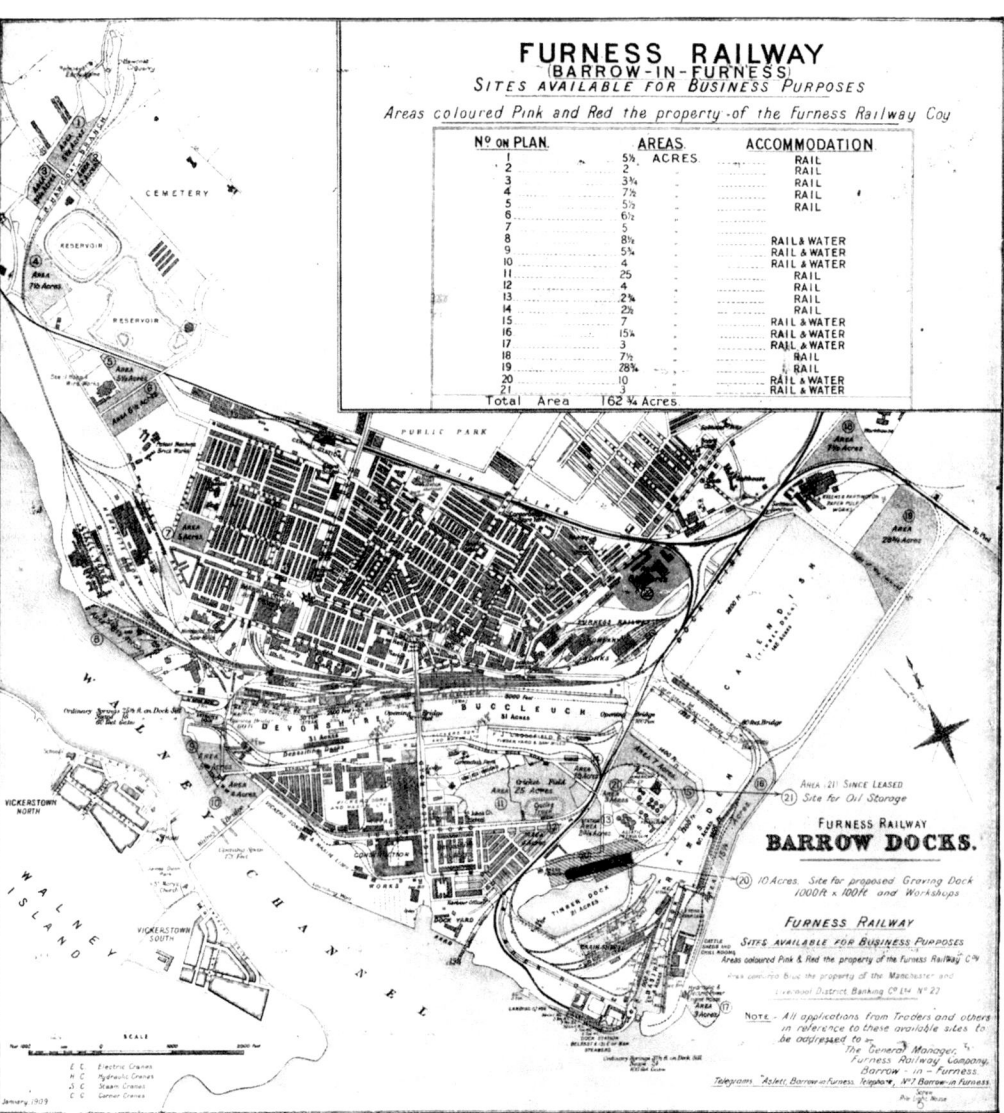

Plan of Barrow Docks produced in 1909 by the Furness Railway to advertise plots of land available for industrial and business use.
Oakwood Press

Paddle steamer *Lady Evelyn* was built in 1900 at Kinghorn, Fife, for the Furness Railway, and remained on their Barrow to Fleetwood service until the outbreak of the First World War when she was requisitioned by the Admiralty. After the war the ship was sold and passed through a number of owners. The ship was requisitioned again for war service in 1939 and sank at Dunkirk after hitting a wreck. *Oakwood Press*

Swift leaving the Furness Railway's Windermere Lakeside railway terminus. Still docked is the distinctively prowed *Tern*. *Swift* was the longest vessel to sail on Windermere, built by T.B. Seath & Co. of Rutherglen then shipped in pieces by rail to the lake and reassembled. She continued in service until 1981 after which she was laid up and used as an exhibition space, before being scrapped in 1997. *Tern* still plies the waters of Lake Windermere, although her steam engines were replaced with diesel ones in the winter of 1957/58.

Oakwood Press

industry to compete with such longer-developed ports as Hull, Liverpool, and the Port of London itself. Local industry was comparatively insignificant; apart from Barrow itself, the rest of the area is purely agricultural. These Barrow Docks were the dream of Sir James Ramsden, and although he defended their construction throughout, no one was more bitterly disappointed than he when, after all the investment and publicity lavished upon them, they finally proved to be a failure, even during his lifetime.

Passenger services were operated from the docks at a very early period, firstly to Fleetwood and the Isle of Man, later to America, as mentioned earlier. The services across the Irish Sea were a joint venture of the Furness and Midland Railway Companies, though the latter was never more than an un interested partner. After the opening of Heysham Harbour in 1904, the Midland transferred its Manx and Irish services thither, which proved a still greater blow to Barrow.

The Furness Railway retained, however, some of its interest in water transport. On Lake Coniston a small steamer named the *Gondola* was put into commission in 1859, designed and probably patented by Mr Ramsden. A second vessel, named *Lady of the Lake*, joined the *Gondola* in 1908. In 1872 the Furness Railway acquired the interests of the Windermere United Steam Yacht Company, formerly two small concerns which had merged; their vessels were the *Swan*, built with Furness Railway assistance in 1870, and the *Raven* of 1871. Later additions to the Windermere fleet included the *Cygnet* and *Teal*, both built in 1879, the *Tern* built in 1891, and the *Swift* built in 1900. There was also a small private-parties vessel known as the *Britannia*, and corresponding to the private saloon carriages on the railway, but further details of the vessel are not known.

Returning to Barrow itself, it was 1854 before the Railway Company had any estate of its own. The growth of the town was not so rapid as Ramsden had anticipated; as yet there was practically no local industry, and apart from the works being carried out on the railway and docks project, there was no attraction for a large number of families to settle in the district. Not until the early sixties, when some industries had been established, did Barrow begin to grow and Ramsden's dream begin to materialise. Thereafter, growth of the town remained steady.

Henry William Schneider, who arrived in the Furness district in 1839, stayed in the newly developing area and went into partnership with Robert Hannay, of Kirkcudbright, in 1853. The firm of Schneider & Hannay was to become a household word in Barrow, and indeed throughout the North. Schneider was a metallurgist of some note, and having been connected with ore mining in different parts of the world, had come to Furness to prospect for iron-ore. First explorations did not prove successful, but in 1853 he took the lease of an old mine at Park, near Askam, which at first excavation had not promised to be of much worth and had been allowed to fall into decay. Within a few weeks it had been established that Park Mine was one of the richest in the district, and the newly-founded firm had made a good beginning. A further small mine was taken over near Lindal in the following year, and from this development grew Schneider's interest in railways, as a means to further his mine production. Both partners then became considerable investors in the Furness Railway.

Toward the end of 1858, Ramsden suggested to Schneider that he should take a plot of land at Hindpool and there build an ironworks. The proposal was welcomed and the partnership sought the aid of one J. T. Smith, who had had considerable experience in ironworks in Staffordshire. The formal opening took place on 17th October, 1859. The new works were not, however, the first blast furnaces in the area, for in 1841 the Whitehaven Iron Company had been formed, and the two West Cumberland Iron Company furnaces were in production in 1854. Schneider & Hannay's works were the first really modern plant in North Lancashire, in comparison with the contemporary charcoal smelting in the bloomeries around Lindal Moor and Backbarrow, and at Leighton,

The Barrow Hematite Steel Company iron works at Hindpool, 1910.

Oakwood Press

near Carnforth. The new Barrow works marked the beginning of a period of prosperity and rapid expansion in Barrow. Within the next few years, five further furnaces were added to the works, and in 1865 the complete enterprise, including the Park and Lindal mines, was sold to a new concern, the Barrow Haematite Steel Company, for £500,000. There were seven directors, Lord Frederick Cavendish (nephew of the Duke of Devonshire), H. W. Schneider, R. Hannay, James Ramsden, the Duke of Devonshire, W. Currey, and F. I. Nicholl, four of these being also Furness Railway directors. It had been found that the haematite ore of Furness, containing as it did no trace of phosphorus, was particularly suitable for the production of steel by the Bessemer process. Consequently, J. T. Smith was instructed to build a Bessemer steel plant adjacent to the works, also a rolling mill for the making of steel rails and other similar products. The main production of the works was steel for rails and railway tyres, and before long Barrow rails were exported all over the world through the docks. From 1867 to 1869 more than 15,000 tons of rails were exported to America alone, and thereby the company gained a strong foothold in the export market. The success of the Barrow Haematite Steel Company led other financiers to invest in ironworks in the area; the next was the Carnforth Haematite Iron Co., of 1866, with three furnaces near Carnforth station. Wakefield, McKinnon & Co. followed, having constructed two furnaces at Askam, whilst the Cumberland Iron Mining & Smelting Co. established a similar works at Millom, on the opposite shore of the Duddon estuary. The latter works had the advantage of the extensive Hodbarrow Mine at its disposal. Thus began the real industrialisation of the Furness area.

A mill and warehouse for the processing of jute and the production of hessian was established in 1870, while later in the same year, S. J. Claye, of Long Eaton in Derbyshire, purchased seven acres of land adjacent to the Furness Railway workshops at Salthouse, where he began to build railway wagons, as a subsidiary of his main works. For quite a lengthy period all ironwork for the wagons was brought from Long Eaton, but a small forge was later established to produce this at Salthouse. The wagon building prospered for some twelve years, whereas the jute works did not; after a precarious existence, including two disastrous fires, one in 1876 and the other in 1892, it was closed. Several other small engineering concerns were founded, and an important adjunct to the shipyards, a ropeworks, started production in 1871. Claye's wagon works, after a very useful period, was hit by labour disputes, and in 1882 the firm was declared bankrupt. A new company, the Patent Tubular Frame Wagon Co., was formed to take over the assets of Claye's works but led a precarious existence since, from all accounts, the railways fought shy of such innovations as tubular framed wagons, and the firm was forced to close down in the late 1890s. After lying derelict for some years, Barrow Corporation eventually purchased the site for building purposes.

Coincident with the opening of the docks in 1867, Barrow-in-Furness was granted a Charter of Incorporation as a municipal borough, with a population of around 16,000. The first mayor was James Ramsden; the council consisted of four aldermen, of whom Ramsden ranked as one, and twelve councillors. The other aldermen were W. H. Schneider, R. Hannay, and Miles Kennedy, representing the most important industries of the borough. Ten of the councillors were also connected, directly or indirectly, with either the railway or the steelworks. Thus it came about that the Furness Railway and the Barrow Haematite Steel Co. enjoyed the most influence in the affairs of the town. Ramsden's plan for Barrow was based on a main street running from the Hindpool Estate and Greengate into the original Barrow village; on either side of this the grid-iron rectangular development was to grow. A small area, including Dalton Road, would not fit into this plan, however, and was developed radially to fill the gap. Apart from the fact that the general manager of the Furness Railway was also mayor of the town, the railway also owned the Town Hall, the Barrow Gas Company, the waterworks, and most of the other amenities; it can thus be said that the railway controlled the town, although it did not afford the accommodation for the town's residents.

Looking north-west from Cavendish Square along Duke Street towards Hindpool. The Town Hall is on the left. *Oakwood Press*

Chapter Two

The Furness Railway

The first proposals for railways in the Furness district were made in 1836, when George Stephenson was contacted and asked to survey a line along the coast. He travelled the whole length of the coastline around Morecambe Bay and up to Whitehaven on horseback. Motivating forces for this railway came both from Ulverston and Whitehaven. Stephenson recommended the construction of embankments across Morecambe Bay from Poulton (now Morecambe) to Humphrey Head (near Grange-over-Sands), with similar crossings of the Leven and Duddon estuaries. A similar idea had been mooted earlier by John Wilkinson, ironmaster, who is otherwise famous for having built the first iron bridge, in Shropshire, and for having sailed the first iron boat on the River Winster near Grange-over-Sands. The Ulverston Committee met again in December 1837, and raised a petition to urge the Earl of Burlington (later 7th Duke of Devonshire) and owner of extensive lands in the area, to proffer his support for the scheme. However, Burlington was not entirely convinced, and so his support was not immediately forthcoming.

In 1838, therefore, John Hague, an engineer who had gained great repute for his reclamation of land in the Fen District of East Anglia, was approached. He also surveyed the area and proved extremely enthusiastic. He proposed the formation of the Grand Caledonian, West Cumberland & Furness Railway, to form part of a trunk route between the South and Scotland. Hague's estimates of the costs for the embankments which he proposed were £362,861, and for the whole scheme, £543,000. This plan, however, roused much criticism, and even the promise of some 46,000 acres of reclaimed land, to be sold at £23 per acre to defray the cost of the railway, did not bring forth much support.

Hague's scheme was that on each side of the bay four rows of piles giving the width of the railway were to be driven across, to meet in the middle. Longitudinal timbers were to be laid along the top of the piles, which latter were to be strengthened by cross struts. The rails would be laid along the sleepers as the work progressed. On the seaward side of the piles, it was proposed that sheet piling be driven down into the sand, grooved into each other, to prevent the tide waters from scouring away the sand itself. At the same time, cofferdams, tide gates and bridges would be built in. The level of the railway would be six feet above the highest spring tide. Then carriages would be driven along the structure,

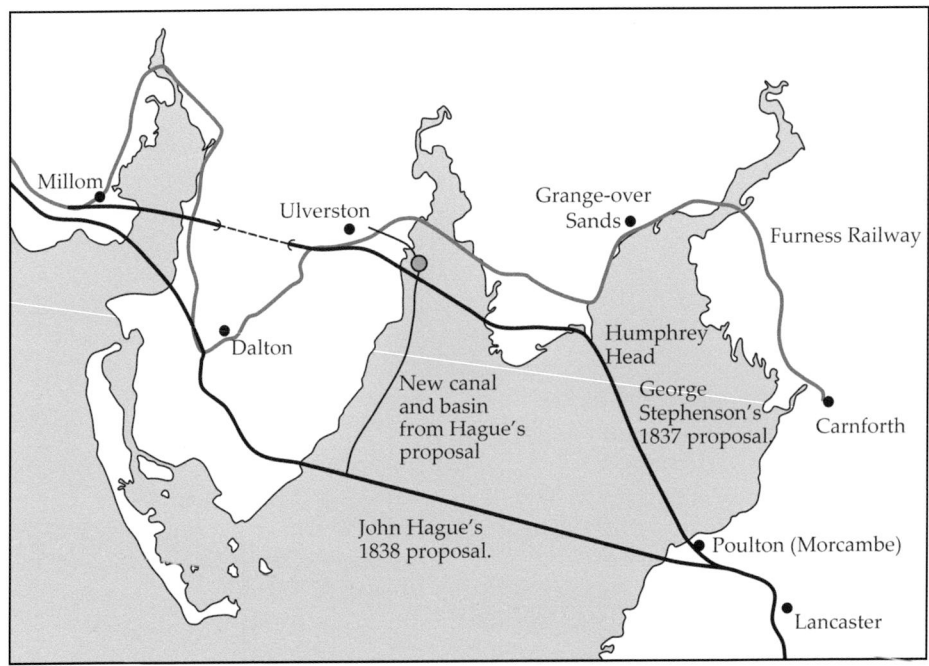

Morecambe Bay schemes showing the proposed embankments across the sands. The Furness Railway as eventually built is shown in grey. As the gull flies it is about 22 miles from Lancaster to Millom, both George Stephenson's and John Hague's proposals are about 23½ miles, and the Furness Railway route is roughly 44 miles.

depositing stones and earthworks scrap from the cuttings into the gaps between the piles, using also the action of the sea to form a shelving beach on the seaward side of the dam, thereby staving off the scouring action of the tide still further.

Two years was the estimated time for the construction of the Morecambe Bay dam, total construction time taking three and a half years. The embankment in Morecambe Bay would need 10,453,785 tons of material for the filling in; 6,149,379 tons thereof would be furnished by the sea through natural tide action. The exact quoted price for the Morecambe Bay dam and Furness earthworks was £289,359 14s. 10d. The works in Duddon were reckoned separately, although the method of construction was the same. Despite the grandiose nature of the scheme, it fell into abeyance.

By 1840 various railways had come as near to Furness as was possible. The Preston & Wyre Railway had reached Fleetwood, which was only some 20 miles distant across the mouth of Morecambe Bay from Walney Island. The Lancaster & Preston Junction Railway was in full operation; parts of the Maryport & Carlisle Railway were also in use. The Preston & Wyre Railway proposed a steamship service across from Fleetwood to Rampside on the Furness coast, with a connecting railway to Ireleth, on the eastern shore of the Duddon Estuary, but this scheme also did not mature. In 1853 a steamship service was inaugurated between Liverpool, Blackpool and Ulverston. It was in connection with these services that John Abel Smith produced his abortive Piel Harbour and Causeway scheme.

When Henry William Schneider arrived in Furness "quite by accident", to quote his own words, and developed an interest in the iron mines there, he was not particularly impressed by the methods of mining carried on there. He found that the workings often flooded due to underground springs, and when water rose above a certain level the workings were abandoned rather than the cost be faced of installing pumping engines. There was hardly a pump in the whole area, and Schneider set about revolutionising the local industry. He obtained several pumps, and having installed these in some mines, under the worst working conditions, had the mines reopened. Whilst prospecting further, he discovered a mine near Askam which had been abandoned due to flooding, the Park Mine. After pumping out, and subsequent thorough investigation, he came to the conclusion that the mine was worth further exploration. A lease was obtained from the Earl of Burlington, upon whose land the mine was, and work began. It soon proved one of the richest haematite mines in the north-west, being computed to contain over half a million tons of fine ore. In conjunction with other ironmasters, Schneider had a new jetty built at Barrow, and to facilitate the shipment of ore endeavoured to get a horse tramway built from the workings to the jetty, but in this he was unsuccessful as the other partners were only faintly interested. Shortage of capital was the principal stumbling block, and Schneider was persuaded to ask the Earl of Burlington for a loan of £40,000. The Earl asked for a guarantee of interest on his money, but as no guarantee was forthcoming the idea was dropped.

Schneider has stated in his diary that "the origin of the Furness Railway was due to pure chance". Both the Duke of Buccleuch and the Earl of Burlington were interested in George Stephenson's original plan for a railway through the district, but then not very keenly. However, in 1841, John Walker was asked to examine the Duddon Estuary and "various railway plans". The Earl owned extensive slate quarries near

Kirkby-in-Furness, and through this was interested in the possible construction of railways as a method of cheap transport for his slate. The shipping trade of Ireleth, where the slate was loaded for transport by sea, had declined considerably between 1834 and 1839, and there was a glut of slate at the quarries in consequence.

Walker completed his report in 1841 and submitted it for the scrutiny of the two peers. In it he suggested that both the noblemen, and the Earl of Lonsdale (who had extensive holdings in West Cumberland) would greatly benefit by any new roads or railways, and that he (Walker) should make a further more detailed survey. The report received moderate acceptance, and Walker was authorised to make his second survey. This report, submitted in 1842, recommended the construction of an embankment on the eastern shore of the Duddon, at an estimated cost of £50,000, which would reclaim 4,000 acres of land, and that a railway be constructed along the embankment from the quarries to Barrow. Burlington commented that the report seemed sensible and that it recommended less than was at first anticipated. Rather strangely, Walker considered that Barrow was the most suitable place for the shipment of the slate rather than Piel or Ulverston, in spite of the difficulties of maintaining the navigability of the Walney Channel.

In the meantime the local ironmasters engaged a surveyor from Kendal, Job Bintley, to go over the area and survey alternative routes from the iron mines near Dalton to Barrow, with the view of constructing a tramway. This was noted by Walker, for in his report he stated that it was to be wondered at that a tramway had not been constructed previously. It would have carried down the produce of the district and would have enabled coal and other supplies to be delivered. It would have benefited all who had property within its influence, and therefore especially the Duke of Buccleuch and the Earl of Burlington and their tenants. A further report to the two noblemen, presented by Walker on 1st June, 1843, detailed a plan for a railway line from Rampside to Kirkby, with a branch to Barrow, plus a horse tramway to the mines at Dalton, in length a total of 18⅜ miles. The total cost was estimated at £100,000. The local ironmasters and merchants received Walker's proposals with mixed feelings; as their landlord, to whom they paid royalties on their iron-ore and rent on their property, was the Duke of Buccleuch, Lord of the Manor of Plain Furness, and one of the interested parties in the proposed railway, they feared that the railway company would be able to extort unreasonable tolls for the carriage of their goods so as to swell the Duke's private coffers. The fears about tolls as expressed may be accepted as perfectly reasonable but were needless, for the Duke was a reasonable man. Stephenson's and Hague's schemes for a trunk railway passing through the area were dismissed, overwhelmed by the feverish activity of the local proposals. The dreams of vast areas of reclaimed land disappeared the same way; from the first plans and arguments of these two ambitious engineers grew a very small railway line.

The prospectus of the Furness Railway Company was published in 1843; in it was expressed the income which the railway might expect to receive, as also the source of this income; from 100,000 tons of iron-ore at 1s. 6d. per ton, £7,500; from 15,000 tons of slate at the same rate, £1,1255; from 25,000 tons of coal and general merchandise, £1,875; lastly, but not least, from the conveyance of 20,000 passengers, £2,000; annual total, £12,500. A reference to the possibility that the line might become part of "any main line of railway which may be extended northward from Fleetwood, and place this area on the highway to all parts of England" was rather obscure, but added a tone of optimism to the prospectus. There was a tentative proposal to extend to Ulverston and for the use of steam traction. This latter may have been designed as a threat to John Abel Smith, whose embankment and pier on Roa Island had just been authorised. Though Piel Harbour had none of the navigational hazards of the Walney Channel and the Barrow Roadstead, construction of the causeway was nevertheless quite difficult, so that the Furness Railway directorate hoped that it had nothing to fear from Smith. Despite this lack of fear, Smith was to prove a thorn in the flesh of the company for the following decade.

The original sponsors of the Furness Railway were the Duke of Buccleuch, the Earl of Burlington, Benjamin Currey (legal adviser to the Earl), Benson Harrison, Montague Ainslie and Richard Roper – the three partners in the iron smelting firm of Harrison & Ainslie; J. I. Nicholl, H. H. Oddie, R. W. Lumley, and F. J. Howard. All of these gentlemen served on the directorate at some time or other, although the Duke of Buccleuch did not become a director for over 20 years, and even after his appointment rarely attended any meetings.

On 23rd May, 1844, the Furness Railway Act received the Royal Assent. There was no opposition to the Bill, this perhaps because the line was purely of a local nature and quite remote from any other railways. The nearest was the Lancaster & Carlisle, still in its infancy, and the Maryport & Carlisle, away to the north, was also just becoming firmly established. Other major companies could recognise no threat to their interests, and therefore offered no opposition.

The authorised capital of the company was £100,000, of which three-quarters was to be issued in shares and the remainder raised by loans. By the time that the Act was passed all the shares had been taken up. Under the chairmanship of Benjamin Currey, the first meeting of the company took place on 17th July, 1844, and was principally occupied with preparations for construction. Subsequent meetings later in the same year agreed to these preparatory proposals. Reference was also made at the later meetings of that year to the fact that a new company had been formed in Whitehaven, under the name of the Whitehaven & Furness Junction Railway, with proposals to construct a line from the West Cumberland town to join the Furness Railway at a point not yet specified. The Furness Railway was not alarmed by the proposals of the new company, as thus far it had no proposals of its own for extension, apart from the tentative one to Ulverston. No threat was seen to the Furness Railway itself; that threat was to materialise later.

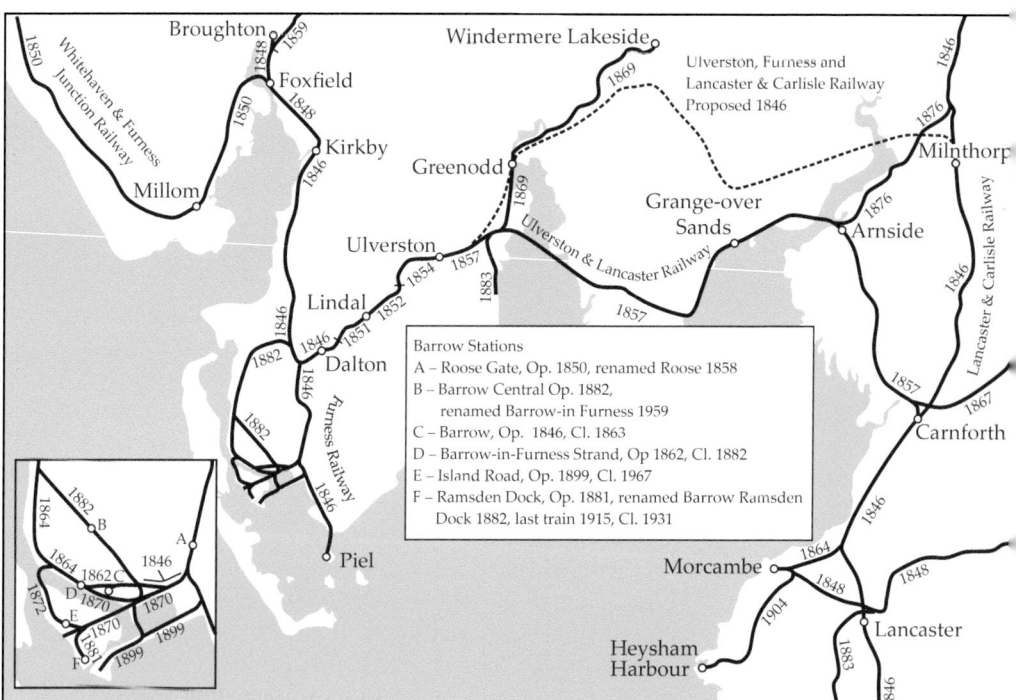

Railway development around Furness and Westmorland. The Barrow railway network is shown on the inset.

The contract for the making of the line was awarded to J. & W. Tredwell, at an estimated cost of £47,789, the track to be completed by 30th September, 1845. By this time, however, the "railway mania" had taken a firm grip on the country, and labour could not be induced to move from the principal lines then being constructed into the remote fastnesses of Furness, which offered little prospect of betterment. The offer of higher wages was therefore made and attracted a number of men, and so the works were commenced. A special meeting of the board was convened on 1st November, 1845, to consider means of raising extra capital to offset the increased cost of labour, this being ultimately done by raising a further loan of £20,000 from the Earl of Burlington.

The route surveyed for the line offered no difficulties, for at no point did it exceed 100 feet above sea level; there were no deep cuttings, nor high embankments to be constructed, few bridges, and no viaducts. James Walker had surveyed his route well. Opposition from the landowners was practically non-existent, in that Buccleuch and Burlington between them owned most of the land, and any other opposition which arose was quickly bought off. The only opposition raised during the construction of the line came from the poet William Wordsworth, who was extremely displeased by the fact that during breaks from work the men engaged on the construction retired to the grounds of Furness Abbey and thereby, in his words, desecrated the area by carousing and playing cards on the tombstones. As the Abbey belonged, however, to the Earl of Burlington, who raised no objection, Wordsworth's opposition was ignored. Shortly afterwards, Wordsworth took an opposite point of view, wherein he expressed his delight at the reverence with which the construction men treated the ruins and the grounds.

As opened, the line consisted of three sections: Kirkby-Dalton, 6 miles 69 chains; Goldmire Junction (just west of Dalton)-Piel, 6 miles 13 chains; and Salthouse Junction-Barrow, 1 mile 27 chains.

In June 1845 the Whitehaven & Furness Junction Railway Company, with the backing of the Earl of Lonsdale, issued its prospectus, and the Bill for the construction of this line was passed on 21st July, 1846. Plans were made by the company for a line from the Furness Railway at Ulverston (providing that the latter built its projected extension to that town) to meet a proposed railway from Leeds to Lancaster (the North Western Railway). About the same time the Lancaster & Carlisle Railway proposed the construction of a line to Ulverston from a junction at Milnthorpe, running through Greenodd, on the shores of the Leven estuary. These tentative proposals became a source of worry to the Earl of Burlington, who noted in his diary that the promoters of these lines were all anxious to gain his support, although he managed to stave off such attempts. It was quite clear that Burlington did not wish to involve himself in any more business transactions, feeling that being occupied with the Furness Railway and Barrow, together with the development of Eastbourne-becoming a popular watering-place at that time-which, was also on land belonging to him, was quite sufficient occupation.

In spite of a little opposition from Burlington, the Furness Railway made proposals for extensions at both ends of the line, from Kirkby to Broughton-in-Furness, and from Dalton to Ulverston; these, embodied in the Furness Railway (Extensions) Act, 1846, received Parliamentary approval in the same year, on 27th July. Another minute of the board of directors proposed either purchasing or building hotels at Kirkby and Furness Abbey, with a view to attracting some tourist traffic. The Furness Abbey Hotel was established in 1847 and expanded in the 1860s connecting it to the new Furness Abbey station. In 1917 following a dispute over licensing arangements with Barrow Council – who refused to grant a license to the hotel's new manager because he didn't live on the premises – the hotel was placed under direct control of the Furness Railway. The Kirkby scheme was allowed to lapse.

In 1846 an agreement was signed with John Abel Smith for the coordination of steamer and train services between Fleetwood, Piel, and Barrow, the steamer service to be operated by Smith.

James Ramsden was appointed locomotive superintendent as from 1st January, 1846, he then being only 23 years of age. Ramsden had served his apprenticeship with, the locomotive-building firm of Bury, Curtis & Kennedy, of Liverpool, and later had worked under Edward Bury at the Wolverton shops of the London & Birmingham Railway. In consequence, on Ramsden's recommendation, Bury supplied the first four locomotives to the company; their livery was Indian red, seen first on the company's lines on construction locomotives supplied by Tulk & Ley, of Whitehaven. The first two of Bury's engines arrived at Piel in March 1846, having been sent by sea from Liverpool. Rolling stock ordered for the opening of the line comprised 60 ore wagons, ten slate trucks, and four passenger coaches, all delivered by sea during the first part of that year. The Dalton-Barrow and Dalton-Piel sections were in use by 3rd June, 1846, although not officially opened until the Kirkby section was ready, on 12th August. Apart from an official dinner, no special ceremonies marked the opening. Ramsden prospered in his position as locomotive superintendent, until he became successively general manager of the Furness Railway, Mayor of Barrow, and director of all public affairs of the district.

THE FURNESS RAILWAY

Dalton station c.1905.　　　　　　　　　　　　　　　　*John Alsop collection*

In 1847 the Furness Railway bought a semi-derelict manor house from the Preston Family, to become the Furness Abbey Hotel. They expanded and connected the building to their Furness Abbey station in the 1860s. During the Second World War it was the headqurters of Barrow's anti aircraft defences and was ironically hit by a bomb in May 1941. It lay in ruins until 1953 when it was demolished.　　　　　　　　　　　　　　*Oakwood Press*

Piel station on Roa Island, the seaward end of John Abel Smith's speculative plan to make money from the railways of Furness by owning the best location for a steamer pier.

John Alsop collection

The causeway from Roa Island made landfall at Rampside station. No. 4 an 'E1' 2-4-0 is arriving with a train from the pier.

John Alsop collection

At the end of the first six months of working, in February 1847, the board proposed a dividend of 4%. It was reported that everything seemed satisfactory and the future should promise even better returns. The pier and jetty at Barrow were purchased from the consortium of ironmasters who owned them, and improved rail and loading facilities were put in, including a primitive type of hopper wagon for the transport of ore. The ironmasters were still annoyed by the freight rates, especially the charge of 1*s*. 6*d*. per ton for iron-ore, but no reduction was granted, for the company held the monopoly of transport in the district. During the winter of 1846-47 it was expected that the newly formed Whitehaven & Furness Junction Railway would make concrete proposals in its plan to link up with the Lancaster & Carlisle Railway, but nothing transpired. The old plan of bridging the Leven estuary beyond Ulverston to Cartmel was also revived, but was not proceeded with, mainly through the opposition of the canal owners of Ulverston and of the shipbuilders and port authority of Greenodd, which would, by the construction of such a viaduct, be cut off from access to the sea. Meanwhile, the contracting firm of Fell & Jopley submitted a tender of £7,200 for the construction of the extension from Kirkby to Broughton, this tender being accepted. The same firm also offered to tender for the extension from Dalton, over Lindal Moor to Ulverston, but this was temporarily left in abeyance.

The line was doubled in 1847 between Dalton and Rampside, the directors having had the foresight to purchase enough land for this at the outset. Seeing a threat of expansion from the company, and his own investments then being amalgamated with the railway, Smith again offered some resistance. He attempted to impose harbour dues on the company for use of the tidal waterways around Barrow and refused to allow them to use Piel Pier. The railway refused the terms offered by Smith and obtained an injunction against him. Early in 1848 they made their own arrangements to take the Fleetwood-Piel steamship service out of his hands. Because of Smith's veto on Piel, the railway company was forced to turn even more towards Barrow, and therefore various improvements were carried out in the roadsteads, including dredging the channels, whereby steamer services could be diverted to Barrow instead of mooring at Piel. Unfortunately, Piel still had the advantage because it was not dependent upon tides, having a deep-water anchorage. Barrow was more tidal and in consequence there was some difficulty in co-ordinating steamship and rail services owing to the fluctuating hours of high tide. Thus Smith retained a very strong position, in control of Piel Roadsteads.

At the half-yearly meeting in February 1848, the dividend was declared at 2%. Overspending was the cause of the low dividend; the extension to Broughton had just been opened but had not sufficiently settled in to be a paying concern. Debenture stock to the value of £25,000 had been issued to date, and a further loan of £6,000 had been procured from the Duke of Buccleuch. Benjamin Currey died suddenly in March 1848, and the Earl of Burlington was co-opted on to the board in his place, being also elected chairman. James Ramsden took over Currey's secretarial duties and was also made general manager. The Earl, in his maiden speech as chairman, said that he had examined the company's accounts in detail and was sorry to report that he found great fault with the management of the railway and with the expenditure, which was far too high. A period of retrenchment would appear necessary. The measures put into force were reduction of staff, postponement of new works, wage reductions, and higher freight charges; needless to say all highly unpopular measures. Despite opposition from some of the board, Burlington was determined either to make the railway a paying concern or to get rid of it altogether. In February 1849 he was prepared to go so far as to offer the whole concern to any major railway which might take it over, but this move was vetoed by the directors. Two months later the panic had subsided, a profit of £3,000 having been made in the previous six months. Construction works on the Dalton-Lindal extension were resumed. On 19th July of the same year, the Whitehaven & Furness Junction Railway opened its first section of line between Corkickle-by-Whitehaven and Ravenglass, from which point a stage-coach service was inaugurated to connect with the Furness Railway at Broughton. Travel further eastwards and southwards could be made by taking train to Dalton, and there connecting with the Over Sands coach to Lancaster or by coach via Cartmel and Levens to Milnthorpe, on the Lancaster & Carlisle Railway. (It might be added that the mails were taken by this latter route as from the opening of the two respective lines of railway concerned.) Alternatively, one could entrain at Broughton, for Barrow, and thence by steamer to Fleetwood.

Burlington was now converted to the idea of continuing the extensions as far as Ulverston, and John Barraclough Fell, who was employed on the construction of the Dalton-Lindal section, suggested another line from Broughton up to Coniston, to tap the trade of the extensive copper mines there, the ore being otherwise moved down Coniston Lake in barges and loaded at the other end into horse-drawn wagons for the continuance of the journey down to the nearest points served by railways. Though the Coniston Railway was not actually proceeded with for a number of years, the board noted the proposals and decreed that henceforth the main consideration of the company should be the carriage of minerals and that passenger traffic should take second place.

In 1846, a Manchester engineer, John Brogden, became a shareholder in the Furness Railway, and five years later he, with his sons John, Alexander, and Henry, produced a plan for a railway from Ulverston to Carnforth, to be known as the Ulverston & Lancaster Railway. The Bill for this line was passed on 24th July, 1851, but little was done in the way of construction work for two years. The Furness Railway regarded the proposed line with some interest as an outlet for their own traffic, but at the same time were very thankful that they would not have to construct the line themselves, for they had had sufficient difficulties with the extension to Ulverston, upon which progress was very slow. Primarily the short tunnel at Dalton had cost much more than was estimated, since the rock formation had proved different to that which preliminary borings had led them to believe; a hidden spring was also causing trouble. The contract for the Lindal-Ulverston section had been let to George Boulton, of Leeds. J. B. Fell had tendered unsuccessfully, and having received this rebuff from the railway company he threw in his lot with John Abel Smith, the latter still enjoying the strength of his position at Piel. The two made a proposal for a direct line from Piel to Lindal, under the name of the Furness & Piel Harbour Company. The

Furness locomotive No.4 leaving the south end of Dalton Tunnel

John Alsop collection

Parliamentary Bill for this undertaking was bitterly opposed by the Furness Railway, backed by the infant Ulverston & Lancaster company, and after a violent argument the Bill was rejected. In December 1852 a severe storm badly damaged the pier and causeway at Piel, and rather than face the costs of reconstruction, Smith and Fell reluctantly agreed to sell their undertaking to the Furness Railway for £15,000. Both then left the district, Fell going to Switzerland to construct railways, and then on to India. Heaving a great sigh of relief that they were at last free from Smith's machinations, the Furness Railway rebuilt both pier and causeway and returned the steamer service to Piel.

Continued bad weather during 1853 delayed construction of the Ulverston extension, and the contractor, Boulton, was reported to be scamping the work. After an acrimonious exchange of views with Boulton, the Furness Railway engineer, McClean, assumed direct supervision of the works. Though only a little over four miles in length, the extension was not opened until April 1854. The Ulverston & Lancaster company also met with some difficulties. The bridging of the Kent and Leven estuaries was the most difficult part of the work. The contract was given to W. & J. Galloway, of Manchester, who introduced a new method of pile driving in the shifting sands, by means of a high pressure water jet. This method was subsequently employed several times afterwards, in the construction of the piers at Southport and Southend, and other similar works. Through the slow progress made on the line, the Brogdens, now resident on Holme Island, near Grange-over-Sands, got into financial difficulties, which in turn put the Furness Railway in a quandary.

Over to the east, in the Pennines, construction was proceeding well on the South Durham & Lancashire Union Railway from Barnard Castle to Tebay, where it was to join the Lancaster & Carlisle Railway. The Furness had a great deal of interest in this new line, as it would give them a short cut for the exchange· of iron-ore for coke with the mines of Durham and North Yorkshire, hitherto this traffic had been shipped round by the Newcastle & Carlisle and Lancaster & Carlisle Railways. Therefore the Furness felt obliged to support the Brogdens, and the Earl of Burlington was asked to loan half the necessary capital, the Duke of Buccleuch having agreed to provide the other half. Pressure on the Brogdens having been relieved, work proceeded satisfactorily, and the whole of the 19 miles 35 chains from Ulverston to the junction with the Lancaster & Carlisle Railway at Carnforth was opened, as a single. line for goods only, on 10th August, 1857. Passenger services commenced a month later, intermediate stations being provided at Silverdale, Grange-over-Sands,

THE FURNESS RAILWAY 35

Ulverston station around 1900. The large name board declares that it is Ulverston, Junction for Lakeside, Windermere and for Conishead Priory. *Oakwood Press*

The viaduct over the River Kent estuary at Arnside. The pier-like nature of the viaduct clearly visable, these piers would later be encased in concrete making the structure much less delicate in appearance. *Oakwood Press*

Lindal station *c.* 1910, looking north to the London Road bridge.   *John Alsop collection*

The copper mines at Coniston were developed around 1590 by the Company of Mines Royal under a charter from Elizabeth I. The mines were important enough to warrant a group of Furness Railway directors to form the Coniston Railway Company to ship the copper to market. Output began to decline in the 1860s and by 1884 the mines had partially closed as a result of cheap imported ore. By 1896 there were no miners at Coniston. There was a brief revival in 1910 when the Coniston Electrolytic Copper Company began exploiting the spoil heaps but the venture closed in 1914 ending copper production at Coniston.   *Oakwood Press*

Kents Bank, and Cark – the latter undergoing several changes of name during the following decades. The working of the line was undertaken by the Furness Railway, the Ulverston & Lancaster not owning any stock of its own, although plans for the purchase of rolling stock were made, also for carriage sidings and a workshop at Cark. Unfortunately, the opening of the line coincided with one of the periodic recessions in the iron trade, and for the ensuing nine months traffic did not meet expectations. The Furness directorate was asked for a further loan of £100,000; this was granted, providing that the Ulverston & Lancaster company would sell its undertaking to the Furness Railway for 4% preference stock at par. This condition was turned down by the Brogdens, and the Ulverston & Lancaster continued alone, still using Furness motive power and stock. Conditions improved, and the two companies worked in uneasy collaboration for five years.

The Ulverston & Lancaster was the last link in the north-western railway system, there being now a continuous line of railway from Lancaster through Carnforth, Barrow, and Whitehaven to Maryport, whence the Maryport & Carlisle Railway continued the line to the Border City (in addition of course to the direct line over Shap). At that period no less than six companies were involved but by amalgamations these were later reduced to three: London & North Western Railway, which came to control the Lancaster & Carlisle and Cockermouth & Workington Railways, together with the coastal line from Whitehaven to Maryport; the Maryport & Carlisle Railway, and the Furness Railway.

To improve facilities in the mining area around Lindal, the short section between Crooklands and Lindal was relaid as a double line, which also involved the enlargement of Lindal Tunnel. The contract for this was let to Tredwells, at a cost of £25,000. The branch between Salthouse and Barrow was also doubled at the same time.

A group of Furness directors formed the Coniston Railway Company at the end of 1856. Though Furness directors were involved, this was an entirely separate company. Its Act of Incorporation was passed on 10th August, 1857, and the contract for the works was let to Charles Pickles, of Bradford, who unfortunately soon got into difficulties, for although the works proceeded more easily than was expected, the contractor was declared bankrupt in August 1858. The line was then completed by local sub-contractors under Furness Railway supervision. It was 8¾ miles long, and was opened on 18th June, 1859, after being 18 months under construction. The Furness Railway took up £10,000 in shares and guaranteed 24% to the lessees of the Coniston copper mines. Gradients were severe, commencing just after leaving Broughton with a long

Broughton in Furness station. *John Alsop collection*

The Furness Railway's Carnforth engine sheds. *John Alsop collection*

stretch of 1 in 49, and climbing continuously at slightly easier grades to a summit near Torver, which at 345 feet above sea level, was the highest point on the whole system. Though mainly concerned with traffic from the copper mines, this time the Furness Railway endeavoured to capture some of the tourist trade, running special excursions to Coniston. In 1862, the FR purchased the Coniston and Ulverston & Lancaster Railways outright. Traffic was up to expectations, and the company was able to double further sections of the main line from Carnforth to Barrow, though it was not until the autumn of 1863 that this work was completed. Dividends between 4% and 6% were paid.

Three Midland Railway directors visited Barrow in 1862 and put before the Furness board a proposal for a connecting link between their Leeds and Morecambe line and the Furness at Carnforth. In the proposal, the Furness was to find half the cost of the construction, and in turn the Midland would divert all its Irish traffic, both passenger and goods, from its own pier at Morecambe to the docks at Barrow. This stimulated the authorities at Barrow to improve the facilities at that port so as to accommodate the Irish steamers, and therefore on 9th October, 1862, to apply for powers to construct new docks. A memorandum submitted to the Midland Railway a few days later put the Furness Railway's point of view in five clauses: -

(a) The capital for the Wennington-Carnforth line, estimated at £120,000, to be held in equal shares by the two companies.
(b) Both companies to exchange equal running powers between Leeds, Bradford, Barrow, and Coniston.
(c) The Midland was to be prohibited from constructing any new lines west of the Lancaster & Carlisle Railway.
(d) The Midland was to be responsible for the working of the joint line, with the Furness maintaining the track and installations, including staffing at the stations.
(e) All traffic from the Midland Railway to the Lake District and Ireland to be routed over the Furness Railway.

The Midland objected to clause (c) as they had just made a proposal for the construction of a new line from Arnside to Milnthorpe, and by using a short section of the Ulverston & Lancaster Railway, gain access to Kendal. After considerable argument, the Midland agreed to drop the Arnside-Milnthorpe proposal, and the Furness decided to build a new line from Ulverston to Greenodd and Haverthwaite, to give access to Lake Windermere. However, five years elapsed before this line was begun.

In regard to clauses (b) and (e) it might be noted that the Midland commenced to use its running powers to the full in summer, after the construction of the Lakeside branch. By using the "Expedition Curve" at

Leven Junction, through trains from Leeds and Bradford gained access to Windermere without the tedious reversal at Ulverston which was previously necessary.

The Furness & Midland Joint Railway Bill was put before Parliament in March 1863, along with the Barrow Harbour Bill. Seven factions opposed the former, but eventually all but the London & North Western Railway removed their objections. The Barrow Harbour Bill was opposed by the Earl of Lonsdale and the Commissioners of the Port of Lancaster, the former regarding his interests in Whitehaven Harbour and the latter because of its supposed authority of the Barrow waterways. The Furness put forward the case that the new line would make the transport of pig-iron and iron-ore to the Yorkshire steelworks much easier than the present route through Lancaster and Preston, which involved two junctions with a main line and either a transfer to the Midland at Lancaster or a double transfer to the London & North Western and Lancashire & Yorkshire Railways at Preston. The Midland Railway withdrew from the argument, but the London & North Western successfully applied for running powers over the whole of the new line between Carnforth and Wennington; of what profit this might be to them is very doubtful, for these running powers were never exercised. However, the granting of running powers secured the support of the LNWR against the Arnside-Milnthorpe Bill of the Midland, which was rejected. Both the Barrow Harbour and Furness & Midland Joint Bills received the Royal Assent on 22nd June, 1863. The Midland almost immediately lost interest in the joint line, but strong opposition from the Furness and LNWR companies forced them to co-operate.

In the following. year, the Furness Junction proposed to shorten the distance from West Cumberland by constructing a viaduct across the Duddon estuary from Millom to the Furness Railway at Lindal. Alarmed at this proposal, the Furness took the extreme course of proposing a Duddon viaduct of their own from just north of Barrow to Hodbarrow, near Millom, even bringing in materials so as to be ready to begin construction. The rival Bills came before Parliament in March 1865, that of the Whitehaven company being accepted while the Furness Bill was thrown out. All these manoeuvrings were an attempt by the Earl of Lonsdale to force the Furness Railway to purchase the Whitehaven line, moves which proved successful for, in self-defence, the Furness bought out the Whitehaven & Furness Junction Railway completely at an inflated price of 8% on capital. The LNWR looked upon these machinations with distinct approval; in its turn it purchased the Cockermouth & Workington and the Whitehaven Junction Railways,

Passengers at Arnside station. *Oakwood Press*

Iron-ore was discovered at Hodbarrow near Millom in 1856 which led to the development of a harbour at Crab Marsh Point near the appropriately named Borwick Rails (Barrow Crails on older maps). As the volume of ore shipped increased Millom Pier was extended becoming almost half a mile long. The pier declined in the 20th century, last shipment was in 1940, and became derelict in the early 1950s. *Oakwood Press*

Viaduct over the River Leven at Greenodd. The village was a port shipping copper, limestone and gunpowder until the construction of the Ulverston & Lancaster Railway's Leven viaduct closed the navigation on the river. *Oakwood Press*

thus also gaining a foothold in West Cumberland. By the schemes of Lord Lonsdale the Furness Railway had to accept responsibility for some 30 miles of poorly-maintained track and to extend its activities to Whitehaven, in addition to which it was obliged to build a viaduct over the Duddon, a viaduct which it did not particularly want.

An agreement signed with the LNWR in 1865 for the construction of a branch from Arnside to Hincaster Junction, to be known as the Furness & Lancaster & Carlisle Union Railway, was in exchange for an undertaking that the LNWR would not construct any lines west of the Lancaster & Carlisle main line. This branch would be of benefit to both companies, since in the first place it would give the Furness a direct route for its coke and iron-ore trains from the North Eastern Railway at Tebay, and secondly the LNWR gained an interest in this traffic, since it owned the intermediate section between Tebay and Hincaster Junction. The Furness was granted running powers from Hincaster to Kendal and Tebay. However, construction of the line did not begin until 1871, and it was not opened until 1876, one year being needed for each mile of its length.

After 20 years as general manager, James Ramsden was made managing director of the Furness Railway, and in the following year the Duke of Buccleuch joined the board. The former secretary of the Whitehaven company, Henry Cook, was transferred to Barrow and appointed secretary of the Furness Railway.

In October 1865, the company applied for permission to extend the Greenodd line through to Newby Bridge, the lowest point on the River Leven at which steamers could call. Steamer services had been in operation on the lake since 1845, under the auspices of the Lake Windermere Steam Yacht Company, their first vessel being named *Lady of the Lake*. Until 1871 there were two rival companies operating services on Windermere, but in that year the Furness Railway bought out the interests of both. In the same year the *Esperance* was launched; she was the private property of Mr Schneider of Barrow; the vessel was one of the first twin-screw ships to be built. H. W. Schneider resided at Belsfield, a large house near the shores of the lake, and he used the vessel to reach the railway terminus, where a special train was laid on to take him to Barrow. Later the *Esperance* became a houseboat.

Also in 1865, the Furness board contacted the Whitehaven, Cleator & Egremont Railway with a view to a connection between Cleator Moor and Sellafield, to facilitate traffic between the coalfield behind Whitehaven and the W&FJR. During this period a large proportion of the Whitehaven-Broughton section was relaid with the Haematite Steel Company's rails; in fact over £50,000 was spent in 1866 on new works and improvements: the Newby Bridge extension, dock warehouses, one new steamer, short extensions at Barrow and Ulverston, and on the Duddon viaduct. About three-quarters of the revenue in 1866-67 was directed to the paying off of debentures, but the directors were quite

Haverthwaite Tunnel. *John Alsop collection*

In spite of many proposals to create a more direct route between Barrow and Millom over the Duddon Sands this original 1850 bridge remained the only railway crossing over the river.
John Alsop collection

Arkholm station on the Furness and Midland Railway's joint line between Carnforth and Wennington.
John Alsop collection

satisfied. The Duke of Devonshire stated that after everything possible had been charged to revenue, the company would still be able to pay a dividend of 10% and would increase its balance by about £1,000.

It was decided to proceed with the Duddon viaduct, since the Whitehaven traffic was increasing, and the price of iron was low. In the previous year, when several new furnaces had been erected, 140,000 tons of ore had been shipped by rail and sea from Hodbarrow, partly from the jetty at that place and the rest through Whitehaven. The Furness Railway directorate hoped that much of this traffic would be carried over the Duddon viaduct to be shipped from Barrow. In contradiction to this, however, a Bill was presented to Parliament in 1869 for abandonment of the viaduct scheme. The argument now prevailing was that it would be uneconomical, owing to the high initial costs, and would involve an annual loss of £6,000. Parliament was not impressed with this argument, and in granting the nullification of the scheme inserted a clause where rates for freight and passenger fares must be made from Millom on the basis of the distance of the direct crossing and not on the actual distance by rail. Thus for many years the third class fare was fixed at 4*d.*, the distance between Millom and Barrow being reckoned as 7 miles, instead of the actual 16¾.

The Carnforth-Wennington joint line was opened on 6th June, 1867, and new docks were brought into use at Barrow in the following August. No sooner had these two undertakings begun to function than a further trade recession, which was to affect the district for the next 50 years, off and on, set in. This recession also caused the cancellation of all new works temporarily. The Wennington line had cost more than had been estimated owing to the high price which had to be paid for the land required; the land alone had cost £60,000 instead of the estimated £10,000. The Furness also tried to effect a temporary cancellation of the Arnside-Hincaster branch because of this trade recession, but were forced by the LNWR, and the coke interests in Durham, to proceed with the line. The Newby Bridge extension, which was actually continued through to a landing stage at Lakeside, was opened in June 1869. One item of satisfaction which emerged from this period of depression was that all outstanding details and differences of opinion on the operation of the Carnforth-Wennington line were amicably settled with the Midland Railway. A purely mineral branch from Crooklands to Stainton, 1½ miles long, came into operation late in 1868. This branch tapped a large ironstone deposit which was opened up to supply the Barrow steelworks.

Carriage of passengers was still of secondary importance at this time. In 1868 out of a total income of £132,000 only £22,000 came from passenger traffic, but as the population of Furness, and of Barrow in particular, increased, passenger receipts rose by 15% in the ensuing seven years. Although some attempt was made to cater for tourists, the general passenger services offered were not good; coaches were mostly four-wheeled and furnished with Spartan simplicity. Compartment seats were narrow, although well backed; the horsehair cloth used to upholster the seats was of poor quality and made for uncomfortable riding, apart from the fact that its slippery nature soon joggled the unwary passenger onto the floor. The LNWR refused to co-operate in regard to train timings, which resulted in poor connections at Carnforth. Only the Irish and Manx boat trains offered some comfort; these trains of Midland stock were handed over to the Furness Railway at a special exchange station at Carnforth, on the direct connecting line between the two companies' metals.

A fleet of paddle steamers, jointly owned by the Midland, Furness, and James Little & Co. of Glasgow, began services between Barrow, Douglas and Belfast in 1867, but the first two years showed little patronage; a gradual improvement took place after 1869.

The year 1874 saw the beginning of the lowest point of the trade depression, which held a firm grip on the area for the ensuing 20 years. Trade in Bessemer steel decreased and the profits of the Barrow Haematite Steel Co. fell sharply. Since the Furness Railway held a large part of the shares of the steel company, its profits also reflected the falling market. In June 1874, James Ramsden reported to the directors that traffic had dropped by a considerable amount and the dividend would have to be cut to 6%. The position was somewhat alleviated by a drop in price of locomotive coal, but it was not sufficient to make much difference. Since the railway held shares in various industrial concerns in Barrow, all of which were to some extent affected by the depression, the bad financial position of the company was magnified. The price of haematite ore fell by nearly one-third in 1875-76 as a result of a surplus being held in stock all over the country. Both railway and steelworks were sufficiently bolstered in capital to be able to weather this storm, but some of the smaller concerns suffered badly. The jute works struggled on precariously until 1877, when its trade bettered itself, but the shipyards took a beating, until the Duke of Devonshire was persuaded to grant them more capital to save them from liquidation. J. T. Smith, the steelworks manager, introduced various severe economies, but the works were affected by strikes in consequence, whilst wage disputes also disrupted Claye's wagon works and even spread to the Furness Railway workshops. The invention in 1876 of S. G. Thomas's "basic" method of producing steel much cheaper than the Bessemer process, which latter needed expensive plant, sounded the doom of the haematite mines. This new process did not need a phosphorus-free ore like haematite, and the traffic in this ore duly declined, until by 1904 the output of the Furness mine was less than one-third of the total production of 20 years previously. Use of this basic method meant that the steelworks were no longer dependent on Furness ore, and the cross-country traffic declined accordingly. In addition, a high-grade iron-ore could be imported cheaply from Spain and at a competitive price, some even coming in through Barrow Docks.

In 1874, the North Lonsdale Ironworks was established near Ulverston, about a mile south-west of Plumpton Junction, to which works the Furness Railway built a branch. Soon afterwards this line was extended to Conishead Priory, on the coast near the entrance to the Ulverston Canal. A passenger service operated sporadically on this branch to the Priory, but in 1912 this was reduced to special excursions only and ceased altogether on the outbreak of war in 1914. Three years later the track beyond the ironworks was dismantled. At the time this branch was mooted as the beginning of an easier route to Barrow, which would avoid the climb over Lindal Bank, and which would also put the coastal villages of Bardsea and Aldingham, which were otherwise isolated, on the railway network. This second line, which would duly take most of the traffic from the original, was to have joined the old

General offices and works of the Furness Railway in Barrow. *Oakwood Press*

route near Furness Abbey. In comparison with the older route, it was not beset by the severe ruling gradients which taxed the locomotives and which made the provision of banking engines necessary.

The Arnside – Hincaster branch opened on 26th June, 1876; three factors played a large role in the slow construction of this line: the recession in trade, the difficulty of obtaining labour, and very bad weather, especially in the winters, three of which were extremely severe. The line was 5¼ miles in length and had two major earthworks, one an attractive viaduct with long approach embankments from Sandside, and the second a very deep cutting to the north of Heversham, which had previously intended to be a tunnel. Intermediate stations were at Sandside and Heversham; the passenger service consisted of three trains daily in each direction, so as not to interfere with the large number of iron-ore and coke trains which worked over this line from Tebay. At the same time that the branch was brought into use, a regular train service from Grange to Morecambe and Lancaster commenced, with requisite running powers granted by the London & North Western Railway.

After various arguments, the Furness and London & North Western Railways took over the Whitehaven, Cleator & Egremont Railway jointly in 1878, by which agreement the locomotive stock of this concern was also divided jointly between the two companies.

Sandside station on the Arnside to Hincaster Junction line. The Furness Railway had running powers over the LNWR from Hincaster to Kendal and Tebay.

John Alsop collection

Chapter Three

The Whitehaven & Furness Junction Railway

In the 1840s the West Cumberland area was considerably more advanced than North Lancashire in railway development, mainly through the enterprise of the Earl of Lonsdale, owner of considerable land and mines in the district, who was very railway-minded. He had been a partner in the outlining of George Stephenson's original grand railway schemes of 1836. Meanwhile, the Maryport & Carlisle Railway had been formed, and construction had commenced; from Whitehaven, the Whitehaven Junction Railway, passing along the coast and through Workington, linked the former town to Maryport. At the same time, the Cockermouth & Workington Railway carried the products of the Honister and Borrowdale mines down to the coast at Workington.

George Stephenson again visited West Cumberland in 1844, and pointed out to the Earl of Lonsdale that the land south of Whitehaven was eminently suitable for the building of railways to develop the natural resources of the district. After considerable discussion, a prospectus was issued for a Whitehaven & Furness Junction Railway

A view over the south side of Whitehaven. A train has just left the Bransty Tunnel and is heading south towards Barrow. The line heading off the bottom left of the photograph leads to Preston Street station, the Whitehaven and Furness Junction Railway's first terminus before they built the tunnel to join the LNER at Bransty.

John Alsop collection

Company. Good support for this proposal was shown from the outset, and a Parliamentary Bill for the construction of the line was passed without opposition on 2nd April, 1847. Great enthusiasm for the new line was shown, and the share capital was over-subscribed within a very short time, but a trade slump, together with the aftermath of the "Railway Mania" of 1846, threatened to put an end to the construction. The Maryport & Carlisle Railway, as a representative of the district, had fallen into Hudson's net in 1846, and many subscribers to the Whitehaven & Furness Junction, holding shares in the more northerly company, were directly affected. However, the Earl of Lonsdale was most insistent that the work be continued, and the construction began therefore from the Whitehaven end in the autumn of 1847. The line commenced at Preston Street, which station later became the goods depot after the opening out of Bransty Tunnel. There were no major engineering works on the line, except for the minor viaduct across the estuary of the River Mite at Ravenglass. From Millom it was originally intended that the line be carried across to the Lancashire shore of the Duddon estuary by a viaduct, which scheme however was dropped, due to the high cost of construction, which could not be met from the existing capital. Instead the line was turned northward at Holborn Hill, in Millom parish, and followed the Cumberland shore of the Duddon through Underhill and crossed the river at one of its narrowest points to the west of Foxfield before turning northwards again to form a Y junction, facing into Broughton-in-Furness station, with the Furness Railway. The line was practically level, there being only one short gradient, 12 chains in length, of 1 in 307 between Drigg and Ravenglass, this being the steepest grade. Near Ravenglass, the contractors unearthed the remains of an old Roman fort, and a number of interesting archaeological pieces were found.

The first section to be opened was that from Whitehaven to Ravenglass, whence a coach was used to connect with the Furness Railway at Broughton, the opening taking place on 1st June, 1849. On 19th July, in the next year, the section to Bootle was opened, and the connecting coach used this station as its terminus. On the completion of the whole line, a special train was run from Whitehaven to Broughton on the opening day, 28th October, 1850, conveying directors and officials of the line. At Broughton, the party retired to the Old King's Head Hotel for luncheon, at which they were joined by the Earl of Burlington, Lord William Cavendish, and other directors of the Furness Railway. The first train services consisted of four trains each way on weekdays and two on Sundays.

THE WHITEHAVEN & FURNESS JUNCTION RAILWAY

Ravenglass station, the terminus of the first section opened from Whitehaven in 1849.
John Alsop collection

The Whitehaven and Furness Junction Railway reached the Furness Railway's line in 1850 and was opened on 28th October. Initially the junction faced Broughton-in-Furness and it wouldn't be until 1858 that the chord to Foxfield station (above) was built allowing trains to travel from Whitehaven to Barrow without having to reverse on the Coniston line.
John Alsop collection

At the Whitehaven end, the Whitehaven & Furness Junction Railway station at Preston Street was separated from the Whitehaven Junction Railway's station at Bransty by Hospital Hill. It was proposed that a tunnel should be bored to connect the two stations and thus provide a connection for a through route to Workington. Physical connection existed otherwise between the two companies by a mineral line which passed through the town's market place to the west side of the harbour, where it joined a dock line from Bransty. This connection was never used for passenger trains; its use lay mainly in the transfer of coal and iron-ore. Passengers had to detrain at the two stations and walk over the hill.

Bransty Tunnel, 1,333 yards in length and single line only, was opened in July 1852, construction having taken over two years. The tunnel gave a direct connection between the two railways running into Whitehaven, and in 1854 the companies arranged for the joint use of Bransty for passengers, with Preston Street the terminus for goods traffic. A further agreement arranged for the joint use of rolling stock. A new station, Corkickle, for passengers only, was built by the Furness Junction company at the south end of the tunnel. Further, in 1860, the Lowther Hotel was taken over and converted into joint offices for the two companies.

The first decade proved relatively difficult, in that there was very little traffic to and from the south. The only really remunerative traffic came

Corkickle station, where a locomotive has just emerged from the tunnel.

John Alsop collection

down from the Whitehaven, Cleator & Egremont Railway for shipment from Whitehaven Harbour, this traffic being in the shape of iron-ore and coal. This was handed over to the Whitehaven & Furness Junction Railway at Corkickle, and therefore the company could only claim freightage over approximately one mile of track. The ironworks at Millom had not yet been built, and no further traffic was possible, except to and from the Furness Railway, there being very little of this. As far as passenger traffic was concerned, although there was slight improvement when the line was eventually linked up to the Furness Railway at Broughton, it was still necessary to take the steamer to Fleetwood, or the Over Sands coach to Lancaster, in order to reach Manchester or Liverpool.

Slight improvement was noted in 1858, when for the first time the W&FJR was able to pay a dividend, though only of 1½%. The Whitehaven, Cleator & Egremont Railway had been completed throughout, and increased iron-ore traffic was handed over at Corkickle from the West Cumberland mines; in addition, the opening of the Ulverston & Lancaster Railway in the late summer of 1857 made possible the prospect of through traffic. In 1858 through bookings to London, Birmingham, Liverpool and Manchester from Whitehaven were instituted, mainly by the 11.15 am "express" from Whitehaven, which took 95 minutes to reach Broughton, with seven stops. There was also a mineral train which ran through to South Staffordshire.

The financial position in 1864 had improved still further, and the question of a direct crossing of the Duddon was again raised. As has already been mentioned, this resulted in protracted legal wrangles which eventually culminated in the purchase of the Whitehaven & Furness Junction Railway, together with all property, assets, and "goodwill" by the Furness Railway.

When Bransty Tunnel was first opened, a telegraph wire was laid through it for the better control of movements on the single line, but the apparatus was distinctly unreliable and was a contributory factor to an accident in the tunnel in 1866, wherein the fireman of a W&FJR train for Maryport was killed. The telegraph reportedly had cost over £100 in repairs during 1865-66, and at the time of the annual general meeting in 1867 had again failed. Lord Lonsdale was of the opinion that it should remain out of action as an expensive but useless luxury to the company. It did, in fact, remain out of use for some considerable time, but was then repaired again, after which it worked spasmodically and erratically for the rest of its existence. It was not until the early 1880s, when traffic through the tunnel had increased to such an extent that complete

Eskmeals station looking north across the bridge over the River Esk in 1911.

Oakwood Press

replacement by an efficient apparatus was imperative, that any attempt was made to ameliorate the position.

An engine shed and repair shop were erected by the W&FJR at Preston Street, Whitehaven, and after the agreement of 1855 these were enlarged so as to deal with the locomotives of the Whitehaven Junction Railway also. The W&FJR also had a small shed at Broughton, which housed one goods and one passenger locomotive. The first locomotive superintendent was William Meikle, succeeded in 1864 by James Rose, who retired in 1866. His son, Edwin Rose, was appointed locomotive superintendent at Whitehaven by the Furness Railway. He remained there until 1880, when he was transferred to Moor Row depot, where he remained until his retirement.

From the opening of the Whitehaven, Cleator & Egremont Railway in 1853, there was a very heavy traffic in coal and iron-ore handed over at Corkickle, whence the W&FJR conveyed it to Whitehaven Harbour over the street tracks. This traffic amounted to 300 or 400 chaldrons daily in the early 1860s. During the hours of darkness, the wagons were drawn through the streets by horses, locomotives being allowed only during daylight. Two small 0-4-0 tank engines were obtained specially for this traffic.

Seascale station about 1905 before the platforms were lengthened or the footbridge built.
Oakwood Press

Silecroft station with the 11.21 am train *c.* 1915.
Oakwood Press

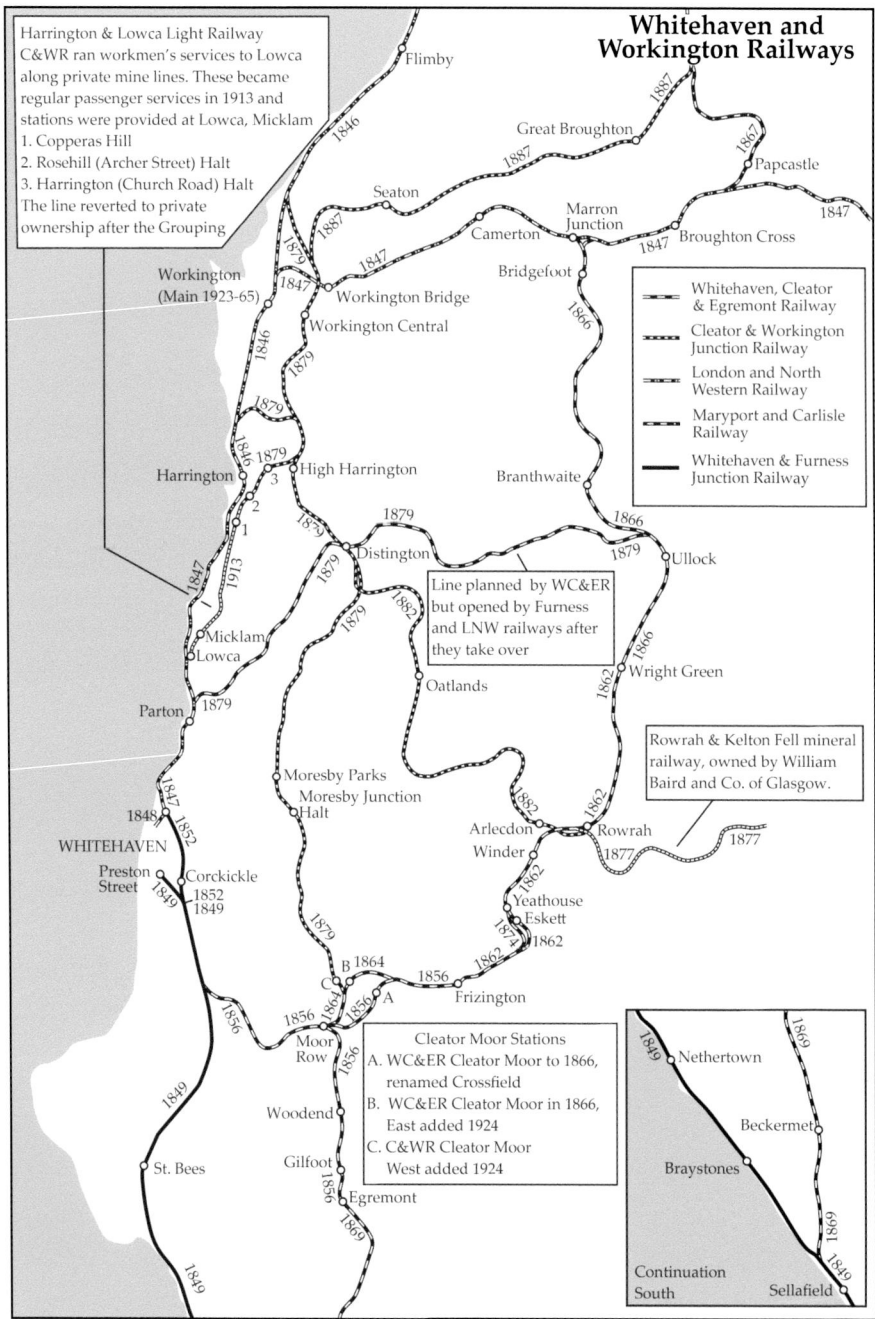

Chapter Four

The Whitehaven, Cleator & Egremont Railway

This was a further company sponsored by Lord Lonsdale and, like the Furness and W&FJR companies, owed its existence to iron-ore. Large deposits of ore had been discovered in the Cleator Moor district of West Cumberland, and until the construction of the railway was carried down to Whitehaven Harbour in horse-drawn carts over very indifferent roads. Lord Lonsdale was instrumental in forming the railway company, which was to take over the transport of coal and ore; it was also this peer who submitted a Bill for its construction to Parliament, and who guided the Bill through, until the company was incorporated in February 1854. Powers were obtained to construct a line from a junction with the Whitehaven & Furness Junction Railway at Mirehouse, one mile south of Corkickle, to Egremont, with a branch from Moor Row to Cleator and Frizington. The line abounded with sharp curves and steep gradients, but nevertheless did not prove difficult to construct, and was opened for goods traffic on 11th January, 1855. Single track was laid throughout, with passing loops. The first dividend declared was 4%, though 10%

could have been paid, the remainder being retained for further extensions and the doubling of the line as far as Moor Row, which was urgently needed as traffic had far outstripped expectations. In the first six months 51,000 tons of ore and 11,000 tons of coal had been carried. The company's engineer, Mr Robson, reported within a few months of opening that the line would soon be ready for the carriage of passengers; it had been inspected by the Board of Trade and pronounced satisfactory, but it was advised that proper coaches be obtained before any services commenced. All stations were completed and ready, but increasing siding accommodation was necessary at Corkickle for the exchange traffic with the W&FJR.

Passenger services were inaugurated early in 1857, with three trains daily in each direction between Egremont or Frizington and Whitehaven. Trains ran through to Bransty station, but the W&FJR imposed a toll of one shillingper train through the tunnel; after a very short time the Cleator company objected to this toll and much argument arose. The W&FJR offered a compromise of sixpence per train, but the Cleator company refused this concession, basing their objections on the fact that as they only had to pay twopence for goods trains, this sum was quite sufficient for passenger trains. The W&FJR remained adamant, and the Cleator company commenced to terminate its trains at Corkickle. As there was seldom a connecting train through the tunnel, this caused great

Staff posing at Egremont station. *John Alsop collection*

indignation among the travelling public, and there ensued much acrimonious correspondence in the local press. Eventually, after considerable hedging by both sides, an agreement was reached whereby all tracks between Mirehouse Junction and Bransty were to be used jointly.

Most of the doubling of the line to Moor Row was completed by the end of 1863. The net profits rose to £9,000 and a dividend of 15% was declared. Construction of an extension from Frizington to Lamplugh, 4 miles, authorised by the WC&ER Extension Act of 7th June, 1861, was well on the way to completion, and a further Extension Act had been passed for the purpose of continuing the line from Lamplugh to a junction with the Workington-Cockermouth line at Marron, the estimated cost of which was £75,000. A deviation at Moor Row, rendered necessary by mining subsidence, was also authorised and commenced.

In 1864 it was proposed to extend the Egremont branch to join the W&FJR line at Sellafield, thus giving an alternative route to the south for the iron-ore traffic and to make this section the joint property of both companies. The Act for this proposal was obtained in June 1864, and in the same month the northern extension to Marron Junction was opened.

Offers were made tentatively by the London & North Western Railway for the taking over of the Cleator company, the attraction being the high dividends, almost always between 10 and 15%, paid by the small company. However, the approaches of Euston met with a sharp

Moor Row station where a LNWR locomotive is working a train of wagons. The engine sheds of the WC&ER and later Furness Railway are in the distance on the left.

John Alsop collection

The Workington, Cleator & Egremont Railway extended their line in 1862 from Eskett to Wright Green station, renamed Lamplugh in 1901. The line was further extended in 1866 to reach the LNWR's Cockermouth and Workington Railway at Marron Junction.
John Alsop collection

Bridgefoot station on the Lamplugh (Wright Green) to Marron Junction line.
Oakwood Press

rebuff. At the general meeting of 1867, the Cleator company chairman, Mr A. B. Seward, remarked that the LNWR shareholders were of the opinion that a good dividend had been paid when it was no more than 5%; the Cleator company preferred to remain independent and earn twice that amount.

During 1868 there was some dissatisfaction over outstanding bills for the carriage of coal and ore owing to the company from certain ironmasters. Some shareholders wanted a special committee established to examine expenditure and general efficiency. This motion was lost after lengthy argument, and a dividend of 8% was declared. Shortly afterwards the Sellafield joint line came into use and provided an alternative route to Barrow for the ore trains.

A general improvement on all sides in 1873 brought a dividend of 12%. However, despite the good rate of profit, the railway directorate decided to increase the rates for the carriage of minerals by 7% for iron-ore and 11 % for coal and coke, which brought forth a storm of protest from the merchants using the line. The company was not unduly worried by the attitude of its customers as it held a monopoly of the traffic in the area. Having been treated with contempt by the LNWR, and now being charged extortionate rates by the Cleator Railway, the merchants discussed a proposal for an entirely new line of railway to run from Cleator Moor through the middle of the West Cumberland coalfield to a junction with the LNWR Maryport line at Siddick, north of Workington, with a branch to the Maryport & Carlisle Railway at Linefoot Junction on that company's Derwent branch. The Bill for this line, the Cleator & Workington Railway, was most strenuously opposed by the Whitehaven, Cleator & Egremont, but in spite of a lengthy hearing, the Bill was given the Royal Assent on 16th March, 1875.

Whilst the ore and coke merchants were thus preparing their revenge on the WC&ER, the latter was applying for powers to construct a mineral branch from Ullock, on the Marron extension, to Distington Ironworks. This, known as the Gilgarron branch, from the name of a large estate through which it passed, was approved by Parliament in March 1875, and in June 1876 further powers were granted to continue the branch from Distington to join the LNWR at Parton, just north of Whitehaven. The aim of this line was to feed the ironworks at Distington with ore from the mines around Lamplugh and Rowrah and also to serve a new colliery which had just been opened at Wythmoor, while the extension to Parton gave an alternative, and shorter, route for the carriage of pig iron to Whitehaven Harbour. In part it was a quid pro quo for the newly authorised Cleator & Workington Railway, construction of which had commenced in 1877. Foreseeing that the Whitehaven, Cleator & Egremont Railway was being subjected to worries from all sides, the London & North Western repeated its offer to take over the company. In a moment of panic, the directors of the Cleator Railway accepted the offer made. However, both parties had ignored the Furness Railway, which company pointed out that the LNWR, by taking over the WC&ER, would be breaking an agreement made some ten years earlier, whereby the LNWR undertook not to penetrate south of Whitehaven and not to build any new lines west of the Lancaster & Carlisle Railway. The Furness also threatened to construct its own line from Seascale to Egremont, Gosforth, and Cleator Moor. The determined attitude of the Furness Railway secured a joint meeting of the three companies at Whitehaven. After much argument, it was agreed that the Whitehaven, Cleator & Egremont Railway be acquired jointly by the Furness and London & North Western companies, which agreement became legal by the passing of the Whitehaven, Cleator & Egremont Railway Vesting Act in 1878. Under the terms of the agreement the LNWR put up the capital of £536,000 and the Furness would be responsible for half the dividend on this amount, fixed at 10% annually, and in perpetuity. Thus disappeared one of the most prosperous smaller companies. At the time of the amalgamation, its stock stood at £170 in the open market, amongst the highest in the land.

A mine locomotive with a train of iron-ore tubs near Birksbridge Junction.

John Alsop collection

Chapter Five

The Furness Railway from 1880

As the amalgamation with the Whitehaven, Cleator & Egremont Railway proved to be the last addition to the Furness system, the main narrative may now be resumed. Almost the first occurrence of 1880 was the decision to build a branch from Askam to Barrow, and to build, within the town limits, a loop line with a new central station, thus putting Barrow itself on the main line. Hitherto it had been at the end of the minor branch passing by Furness Abbey and through Roose. Passengers from Barrow for the north were obliged to travel out to Furness Abbey and there change, or alternatively to travel to Dalton and there board a through train. The direct line between Foxfield Junction and Green Road, avoiding the reversal at Broughton, had been constructed just after the Whitehaven & Furness Junction Railway became Furness property. The new loop, and the Barrow-Askam line, with a total length of 7½ miles, was estimated at £48,000. The line, however, was not easy to construct as, due to the expansion of the town, many under- and over-bridges were necessary. The new premises at Central station were ready for use and were opened for traffic on 1st June, 1882. All passenger trains between Carnforth and Whitehaven were then diverted to this route, the avoiding line (the original main

A train heads east through Askam, c. 1910. *Oakwood Press*

The exterior of Barrow Central station. The glass display case, built in 1907, that held locomotive No. 3 is on the right. The station was almost entirely rebuilt in the 1950s after its destruction during the Barrow Blitz, on 7th May, 1941. *John Alsop collection*

Shed and platforms of Barrow Central station. *John Alsop collection*

line) thereafter being used only by 'goods trains not required to call at Barrow. The original station, on the Strand, was converted to a goods station, this station in its turn having superseded a small wooden building with one platform; a terminal station of which the buffer stops were at right-angles to the run of Barrow Channel.

During the ensuing years the railway suffered from further depressions in trade, and dividends fell from 6½% in 1880 to 2% in 1886. It was proposed in 1883 to raise an additional £200,000 in capital; mineral receipts had fallen by £31,000. Certain shareholders felt that far too much money had been expended, abortively, on the Barrow Docks, and that under the present financial stringency such expenditure could now be termed unjustified. The chairman replied that the only works carried out at the docks were necessary developments. He pointed out further that no complaints had been voiced when dividends were high. In the following year the shareholders' attack on the directorate was renewed with vigour. Trade had not improved, and the board had just announced a proposal to deepen the Walney Channel. One shareholder stated that the docks had already cost the company well over two million pounds and were a millstone round the neck of the railway. A rumour spread that the Furness Railway was to be absorbed by the Midland, but this was vigorously denied by the chairman.

In 1884, one of the periodic violent storms visited the district and, amongst other damage, destroyed three-quarters of a mile of the embankment between Arnside and Sandside, on the Hincaster branch, causing the diversion of the coke and ore trains to and from Tebay, through Carnforth, while passenger services were entirely suspended. Nearly seven months elapsed before normal running over the branch was resumed.

The worst stages of the slump had been passed by 1886, but once more the annual meeting proved stormy. Again the militant group of shareholders returned to the attack, with the same subject, the docks. They alleged that if all the money spent on the docks in the previous ten years had been saved, the company could have paid dividends as high as 20% instead of the meagre 2% which had just been announced. One point which the same shareholders conveniently forgot was that at all times the company had managed to pay some form of dividend, which claim could not be made by other companies, including some of the larger ones. The shareholders did, however, extract an admission from the chairman that overtures had been made to the Midland Railway, and that at one point negotiations had actually commenced. However, the discussions had fallen through because the Midland had not been prepared to guarantee more than 5% on the Furness holdings. Fortunately for both shareholders and the directorate the iron and steel trade improved considerably during the year, and mineral receipts rose by more than £4,000.

The recovery of the company's fortunes seemed assured in 1889 when it was possible to declare a dividend of 5%, but no sooner had good trade figures been established than another slump occurred in the following year. An increase of 2s. 2d. per ton in the price of locomotive coal caused working expenses to rise by £1,050, whilst receipts fell by almost the same amount. The resultant desire to economise made itself felt in the train services, a number of which were decelerated; the up Whitehaven Mail was slowed by ten minutes; the up Midland boat express from Barrow to Carnforth took 50 minutes non-stop, 8 minutes longer than before, and the best timing was taken over by the 3.43 from Carnforth, which needed 2 hours 12 minutes to reach Whitehaven.

In 1891, the Duke of Devonshire died and his place as chairman of the Furness Railway was taken by his son, the Marquis of Hartington, who also succeeded to his father's title, and held the chairmanship till 1908.

The annual general meeting of 1893 was no better than the previous four. An attempt was made to separate the finances of the docks and steelworks from those of the railway. This point was argued at considerable length by the company's auditors, but the finances of the three undertakings were so inextricably involved with one another that such a task became hopeless. The militant group of shareholders demanded that Sir James Ramsden be compulsorily retired from the general managership, as being responsible for the expenditure on the docks. The Duke of Devonshire decided to treat the matter as a vote of confidence, and having put it thus to the meeting, the resolution of confidence was passed by a narrow majority. The question of the Duddon crossing was again raised and gained a fair amount of support, but the chairman decided that the matter had been investigated very thoroughly, and though there would be a considerable saving in train mileage, the construction of the viaduct and the relative costs would still cause an annual loss of £12,000 after completion. Also, any chances of increased traffic from the proposed viaduct were very problematical, and so the plan was again allowed to lapse. It did not appear again during the independence of the Furness Railway, but from time to time since 1923 it has been brought into prominence, though always being shelved after considerable discussion.

With the implementation of the Board of Trade's Bill for the use of continuous automatic brakes on passenger trains, the Furness was

HMS *Dominion* being completed in the Devonshire Dock, January 1905.   *Oakwood Press*

Looking through the Cradle Bridge which crossed where Buccleuch and Ramsden Docks met. A Scherzer rolling lift bridge, a refinement on bascule bridges that allowed quick opening and closing, it was officially opened in 1908 and was removed in 1971 to widen the passage between the docks.   *Oakwood Press*

compelled in 1893 to purchase 55 third class carriages, fitted with the automatic vacuum brake, at a cost of £24,319; the remainder of its rolling stock was converted in Barrow Works.

A strike in the coalfields of Lancashire and Yorkshire dealt a severe blow to the Furness district, and out of 75 blast furnaces only 31 were able to maintain production. This produced soaring unemployment figures and sadly reduced passenger traffic. In addition, the competition of Spanish iron-ore was developing keenly; for example, the Furness Railway carried 268,000 tons of locally mined ore to places beyond Carnforth in 1874, but 20 years later, in 1894, this total had fallen to 27,000 tons, one tenth of the previous figure. Consumption of local ores, and consequently coke and limestone, in the blast furnaces of the district was also greatly reduced.

Happily, the year 1895 saw the turning point in the railway's fortunes. After 20 years of having to struggle, the situation began to improve. Important changes in the directorate helped considerably towards this. Mr Cook, company secretary for the past 30 years, retired and was succeeded by Alfred Aslett, who came from the Cambrian Railways. He was well qualified, for the Cambrian, a railway very much of the nature of the Furness, was a company which depended largely on passenger traffic. In the ten years after his appointment, Aslett changed the outlook of the Furness Railway completely; instead of being dependent on mineral traffic, this took a second place to passenger services. In 1896, Sir James Ramsden died, and Aslett became general manager, retaining

Hodbarrow No. 6 iron mine. The Hodbarrow mines were established to exploit the largest haematite ore body found in Britain. *Oakwood Press*

the secretaryship also. He was foresighted enough to realise the undeveloped facilities offered by the Lake District and that these should be exploited. Apart from the Kendal & Windermere section of the LNWR, and the Cockermouth, Keswick & Penrith Railway in the north, the Furness held a virtual monopoly of rail services into Lakeland; from operating its through trains along two sides of a square containing the Lake District, it had access to the third side from Workington to Keswick and, by means of the Hincaster branch, to part of the fourth side. In addition it had its own branches into Lakeland via Ulverston and Lakeside, serving the Windermere steamers, and to Coniston. Previous to 1896 only four combined rail and coach tours of the Lakes had been operated, and these only during the summer; the best of the 20 which Aslett immediately organised was the Six Lakes Tour, which covered Windermere, Ullswater, Derwentwater, Thirlmere, Grasmere, and Rydal Water, all for 13s., including steamer fares on Windermere and Ullswater, coach over Kirkstone Pass, rail to Keswick, and coach back to Ambleside. A very popular short tour from Barrow was to Cartmel Priory and Holker Hall, for 4s. 3d. A Sunday service of steamers started on Windermere in 1896 previously had operated only on weekdays: a series of cheap special Sunday excursions was instituted. Another innovation was the cheap weekly ticket, allowing one return journey daily between the two places named on it. Second class was abolished in 1897, except on the West Cumberland lines, since the LNWR still retained it. Further innovations were through trains run by the Midland Railway from Leeds and Bradford to Lakeside, and also from Morecambe Promenade station, whilst the Great Central Railway arranged one or two through carriages from the south to Lakeside, via the Midland Railway.

Another innovation, which ran for some years, was a through carriage from Whitehaven to Southampton, for the use of emigrants. This was worked forward by the Midland Railway from Carnforth to Cheltenham, whence it left attached to the 5.00 am American & Cape Lines Boat Express of the Midland & South Western Junction Railway. This service, run once a week, connected with a special Midland train from Bradford, usually run on Friday nights, but departing sometimes twice or even three times a week from the northern city.

The Naval Construction & Armaments Co. obtained a lease of part of the foreshore at Eskmeals, for the purpose of constructing a gun-testing range; a special halt station was built by the Furness and through special trains from Barrow were run, this also bringing extra passenger traffic to the railway.

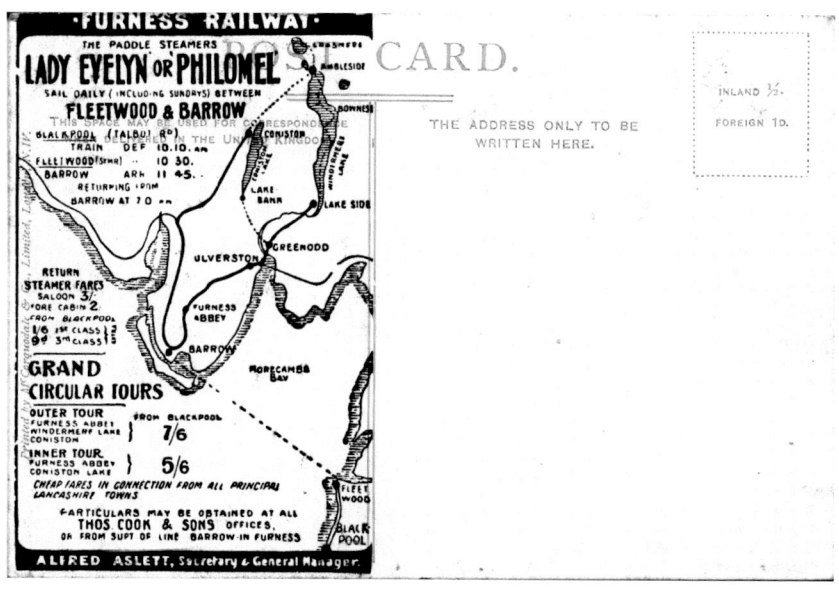

Postcard produced by the Furness Railway to advertise their tours of the lakes from Fleetwood on the PS *Lady Evelyn* and PS *Philomel*.   *John Alsop collection*

Mr R. Mason, who had been locomotive, carriage, and wagon superintendent since 1850, retired in 1896 and was succeeded by W. F. Pettigrew, who came from the London & South Western Railway, where he had worked under Adams. Apart from a manual on steam locomotive construction and working, which became the standard work of its kind, Pettigrew brought a great deal of Adams's influence into the locomotives which he designed for the Furness Railway. In fact, the company's locomotive policy was completely changed, for Mason had had no part in the designing of Furness locomotives; all up to his retirement had been standard productions of the various makers from whom they were obtained. Pettigrew immediately commenced to design his own engines, and though they were all built by outside firms, there were never again any "standard" designs-in the makers' sense-for Pettigrew was a great believer in standardisation with his own designs.

Aslett's policies soon began to take effect. Between 1895 and 1898 passenger traffic increased by 12% and greatly helped to offset the decline in minerals caused by the poor state of the iron trade. During 1896, Barrow Steelworks closed for six months and 4,000 were laid off. By 1898 trade had sufficiently increased again in goods and mineral traffic to allow the railway to make some small profit in this field; goods traffic was up by 17,000 tons and minerals by 55,000 tons. Vickers, meanwhile, as successors to the Naval Construction & Armaments Company, were now employing 7,000 men, and there was some consequent increase in passenger traffic.

About this period, 1897-1900, many of the older locomotives were wearing out and were withdrawn or sold. The new locomotive superintendent designed stock to take their place. Most of the main line was also relaid with 99 lb. rails, and some experimental lengths of 100 lb. flat-bottomed rails were put in between Ulverston and Barrow. Aslett also arranged for through carriages, to run throughout the year, from London (Euston and St. Pancras), Liverpool, Manchester, Leeds, Bradford, and Sheffield to Barrow and Whitehaven.

Further steps were taken by Aslett to develop the steamer services between Liverpool, Blackpool, and Barrow, thereby attracting holidaymakers to take day trips by steamer, rail, and coach to the Lake District. The first vessel for the service, a paddle steamer named *Lady Evelyn*, was 200 ft long by 24 ft beam, and had a speed of 17 knots. During the summer of 1903 over 41,000 passengers were carried by this vessel and a sister ship, *Gwalia* (later renamed *Lady Moyra*); a third ship, the *Lady Margaret*, had been built in 1903. All three were named after

The PS *Duchess of Buccleuch* of the Barrow Steam Navigation compay – a partnership between James Little & Company of Glasgow and the Midland and Furness Railways – at Ramsden Dock station. From there the ship carried passengers to the Isle of Man and Belfast.
John Alsop collection

members of the Cavendish family, of Holker Hall, of which the Duke of Devonshire was the head.

The Midland Railway caused some consternation in 1904 by announcing that consequent upon the opening of a new harbour at Heysham, it would transfer all its Manx and Irish traffic to that port. Though the route through Barrow involved a longer train journey, the sea voyage was considerably shorter. Protracted negotiations with the Midland produced an agreement whereby the latter would maintain a service through Barrow to Ireland for three years, and to the Isle of Man for two years, as from 1905. The Midland also purchased the Furness interests, as too those of James Little & Co., in the steamer services for £45,000, and guaranteed to route a portion of the traffic to Belfast via Barrow. Heysham Harbour was officially opened on 1st September, 1904.

To offset the loss from the Midland traffic, an agreement was entered into with the North Eastern Railway to run a through service from Newcastle-upon-Tyne to Barrow, via Darlington and Tebay, to connect with the steamers to and from the Isle of Man. The train left Newcastle at 9.30 am and arrived at Barrow at 2.10 pm, to connect with the afternoon steamer to Douglas. The return train left Barrow at 12.15 pm, after the arrival of the morning boat, and reached Newcastle at 5.07 pm A new combined rail, sea, lake, and coach tour was announced from Blackpool via Fleetwood, Barrow, Lakeside, Ambleside, Coniston, with return through Barrow to Blackpool, for 7s. 6d. Bad weather during the summer of 1907, plus the recurrent depression in the iron and steel trade, made this a bad year, with passenger traffic making a very small profit, but, with goods and minerals, down by just under £1,000.

At the Franco-British Exhibition of 1908, the Furness Railway rented a stand so as to advertise its services through the Lake District. Special folders, printed in French, were distributed, an entirely novel departure from the usual forms of railway advertising. The adventure proved rewarding and had considerable success in attracting foreign tourists over the area. However, despite this, dividends fell to the lowest ever in 1908, being only ¾%. The Duke of Devonshire died in this year and his place as chairman was taken by his nephew, who had been a member of the directorate since 1890. Contemporarily there were various changes in the executive. The company's engineer, W. S. Whitworth, retired through ill-health, and was replaced by D. L. Rutherford, of the North British Railway, where he had been in that company's engineering department as a junior assistant. Almost immediately the new engineer commenced planning for the reconstruction of all sections of the permanent way, which on completion bore comparison with the best of

THE FURNESS RAILWAY FROM 1880   73

The Furness Railway's stand at the Franco-British Exhibition advertising the beauty of the Lake District.
John Alsop collection

Silverdale station, on the Furness Railway's main line to Carnforth.
Oakwood Press

any railway company's track work in the kingdom. Over 50 bridges and viaducts, including those over the Kent, Leven, Duddon, and Calder, were rebuilt or renewed.

Mr Rutherford was also responsible for the deepening and widening of the Walney Channel and for the provision of a large floating dock, with other improvements to the docks themselves. A new dredger (still at work 50 years later at Heysham) was provided the *Myles Kennedy*. The engineer was a member of the Institute of Civil Engineers, Member of the Institute of Transport, Fellow of the Permanent Way Institution (its president in 1912), and also held the rank of major in the Engineer and Railway Staffs Corps, R.E. (T.A.).

Another change on the executive side saw the appointment of Mr F. J. Ramsden, son of the late Sir James; he had been superintendent of the line since 1890.

A regulation was promulgated by the directorate that all the company's apprentices attend technical classes organised by the Barrow Education Committee and that all successful candidates be given extra pay. First aid classes were also organised, with sub-divisions at Ulverston, Millom, Whitehaven, and other centres, and a silver shield put up for competition between first aid teams from departments of the company and the docks.

A number of accelerations in train timings were made in 1911. The 4.43 pm from Carnforth reached Whitehaven at 7.10 pm and bore through carriages off the 11.25 am from Euston. A Mondays-only through carriage left Whitehaven at 7.45 am, instead of 6.40 am, but still reached Euston at the same time. Meanwhile, the Midland lost all interest in Barrow, and the through service to St. Pancras was cancelled, although an extra through train was put on from Leeds and Bradford in both directions. At the outbreak of war in 1914, Aslett had improved the passenger service by 102% over the 18 years during which he had been in office; apart from this, gross receipts had increased by 65%.

Cover of a packet produced by the Furness Railway to hold a selection of their postcards.
John Alsop collection

Official railway postcards were used by a railway company for their correspondence, or were given away or sold by the company to promote their routes. Often these are of scenic places served by the railway. Some show locomotives and carriages, others are of hotels and tourist attractions owned by the company.

Furness Railway postcards showing Lake District beauty spots together with their most convenient station.

FURNESS RAILWAY OFFICIAL POSTCARDS

For the Furness Railway postcards were an important part of re-creating its image as a tourist line, away from its past as an industrial venture. Tourism to the Lake District had been established, through the eighteenth century writing of Thomas West, and cemented in the nineteenth century by the romantic poems of William Wordsworth and others. The Furness Railway was perfectly placed to capitalise on their legacy.

FURNESS ABBEY HOTEL, HALL & INGLE NOOK.

FURNESS ABBEY HOTEL, COFFEE ROOM
SHOWING BAS-RELIEF "CREATION OF EVE" FORMERLY IN FURNESS ABBEY.

The company's Furness Abbey Hotel postcards showed off the fine accommodation.

THE FURNESS RAILWAY

While the Furness Railway didn't invent the allure of the Lake District it was important in making it available to working people. They provided the means of reaching the Arcadia and once tourists were there they were offered tours of places they had read about, or glimpsed in the company's posters and postcards.

The artist George Romney's (1734-1802) early home was among the Furness Railway's possessions. In 1909 they opened a museum in the cottage, where he had lived from 1742-55. George Romney was the most fashionable portrait painter of his time. He was particularly fond of Emma Hamilton whom he met in 1782, and painted more than 60 portraits of her; she would later become notorious as the mistress of Lord Nelson. The cottage museum had several reproductions of Romney's paintings of her and his other works. It closed during the First World War and never reopened.

Chapter Six

The War Years and the Grouping

The Kaiser's War found the Furness Railway in a very strong position. The area which it served was a very important part of the country from the standpoint of war effort, because of the situation of the steelworks, the naval construction yards, and the armaments plant at Barrow. The town took on a leading role, and for the Furness Railway itself this meant large increases in mineral traffic. After the first year of war, however, the cost of wagon repairs had risen by £10,000, and partly to offset this the increase in workmen's tickets by over 3,000 reduced the costs somewhat. A number of extra carriages had to be provided to cope with the workmen's traffic. In 1916, the Marquis of Hartington, then having succeeded to the title of Duke of Devonshire, resigned his chairmanship of the board on being appointed a Civil Lord of the Admiralty. Lord Muncaster then took his place but resigned the following year, and was succeeded by the deputy chairman, F. J. Ramsden. Goods and mineral traffic increased to just under 5 million tons in 1916 and a dividend of 24% was declared, which rate was maintained until the end of the war.

There was a chronic shortage of wagons and locomotives; large requisitions of rolling stock were satisfied from South Wales, while the London & North Western, Midland, and North Eastern Railways provided locomotives "for the duration", and the Maryport & Carlisle worked some trains on behalf of the Furness. LNWR engines worked as far as Lindal sidings from Carnforth; from Whitehaven one of the principal passenger trains, the 7.05 pm. Mail, was worked as far as Millom by a M&CR engine, which returned to its own territory on the last down passenger train of the day. It was also arranged that the morning down fast goods from Carnforth worked through to Workington, thereby avoiding the exchange shunts at Corkickle and a separate LNWR turn from Whitehaven to Workington. Admiralty coal specials for the north of Scotland were worked on Sundays over the Furness and M&CR lines to Carlisle to relieve congestion on the main line over Shap; these trains were usually worked throughout by LNWR "Cauliflower" 0-6-0 engines between Carnforth and Carlisle.

The gun-testing range at Eskmeals was greatly extended, and Vickers laid in an extensive private railway, which was connected to the Furness main line just south of Eskmeals station. A further halt, Monkmoors, was established at the junction for the convenience of Vickers' workmen, with special trains run for them from Barrow and Millom. The block section between Bootle and Silecroft, over five miles long, was divided by the erection of a signal box and crossing loop at Stangrah. A further loop was laid a mile south of Corkickle, between the single Furness line and the double Whitehaven, Cleator & Egremont line, which here ran parallel to each other (though separate) to Corkickle. A new signal box, Corkickle No. 1, was erected, and the Furness single line thence to Mirehouse became a "permissive" goods line, while all passenger trains used the Egremont lines. Several additional miles of sidings were also laid at Barrow to cope with the war traffic.

The increase in the number of heavy munitions trains rendered necessary the reconstruction of the viaducts over the Kent and Leven, lying as they did on the route from the south to Barrow. When the original viaducts were constructed, large quantities of slag and limestone had been dumped beneath the superstructure and surrounding the columns so that a form of causeway had been made from either bank to the central opening or wide spans. In the course of time, the erosive action of salt water had made the iron of the piers so soft that it was possible to hammer heavy nails through them. In the reconstruction, the causeway, which varied in depth from six to ten feet, was excavated to a depth of five feet, and into this was poured a foundation of reinforced concrete. On top of this concrete steel caissons were erected, and the latter filled with concrete, the old iron pillars in the middle of this acting as bearing piles. On top of these caissons a

River Kent viaduct after the piers were encased in concrete. *Oakwood Press*

surrounding casing of brickwork was built, enclosing the separate structures. Thus a much strengthened pier was formed, which together with the upper parts of the viaducts carrying the rails formed bridges capable of meeting the demands of the heaviest trains passing overhead.

Mr Aslett retired in 1918 at the age of 71. His retirement was universally regretted, as he was held in high esteem by both board and employees. Through his efforts the position of the railway and the town and tradespeople of Barrow had been much improved. Mr Pettigrew also retired in the same year, and was replaced by Mr Rutherford, who took over the post of locomotive, carriage, and wagon superintendent, in addition to his duties as civil engineer.

A new dry dock, capable of taking the largest warships, was constructed at Barrow at a cost of £100,000. The trade boom which followed the war occupied the railway fully, but by 1921 a recession again set in. As with other railway companies, a lot of leeway had to be made up in the repair and maintenance of locomotives, rolling stock, and permanent way. The expenses incurred thereby caused the dividend to drop to 1%.

With the imminence of the Grouping, under Sir Eric Geddes' Railway Act of 1921, the Furness Railway, like so many other railway companies in Great Britain, "marked time" during 1921 and 1922, as new developments could not be implemented; all that was done was routine maintenance. In due course, the Furness Railway became a constituent of the London, Midland & Scottish Group.

To end this survey, a glance at the pattern of train services in 1922 might not be irrelevant. On the main line, the first down train of the day was the 4.40 am Mail from Carnforth, which stopped at Grange to drop mails and then, with further stops at Ulverston and Dalton, reached Barrow at 5.36 am, where it stopped for ten minutes. With further stops at Foxfield, Millom, Bootle, and Drigg, it reached Whitehaven at 7.25 am The next train left Carnforth at 6.58 am, and called at all stations except Lindal, Green Road, Braystones, and Nethertown, and arrived at Whitehaven at 10.00 am The 9.35 am down was a slow train, calling at all stations, and arriving at Whitehaven at 1.20 pm, after which the 1.30 pm performed exactly the same service, arriving at its destination at 4.55 pm; this train, however, carried through coaches which had left Euston at 6.55 am The last two trains from Carnforth to Whitehaven were the 4.20 pm and the 7.10 pm The former was a semi-fast, stopping at Grange, Ulverston, Barrow, Millom, and then all stations, and arriving at 7.10 pm The latter of the two through trains gave the fastest down service, and with seven principal and four conditional stops, arrived at

Whitehaven at 9.35 pm On Saturdays only, this train had two additional stops, though no extra time was allowed.

The up service commenced with the 6.35 am from Whitehaven, which, omitting two stations to Seascale, then called at all stations to Carnforth, and conveyed a through carriage for Euston, due to arrive there at 4.30 pm After that came the 10.15 am, calling at all-stations to Carnforth, arriving at 1.43 pm At 11.35 am the best up train of the day left Whitehaven, and calling at St. Bees, Seascale, Millom, Barrow, and Ulverston only, arrived at 1.53 pm, directly after the 10.15. This train also conveyed through coaches for Euston and Leeds. Then, leaving Whitehaven at 1.50 pm with the designation semi-fast, a train called at all stations to Millom, then Barrow and Ulverston only, arriving at Carnforth at 4.38 pm, again with a through carriage for Euston. The 3.05 pm slow was the counterpart of the 6.55 am down, omitting the same stations, and reaching Carnforth at 6.03 pm At 5.30 pm a further all stations left Whitehaven, picking up the mails from the smaller stations, and acting as a feeder for the 7.00 pm up mail, which called at Sellafield, Seascale, Millom, Askam, Barrow, and Ulverston, arriving at Carnforth at 9.22 pm Although the Furness had two mail-sorting carriages, it did not have any pick-up apparatus, this being the reason for the slower train preceding the up mail, to enable the latter to leave somewhat later and to stop only at strategic stations.

Soldiers on guard at Barrow Central. *John Alsop collection*

The intermediate services comprised five trains in each direction between Carnforth and Barrow, with the exception of the 4.25 pm, which ran through to Millom. The first train, consisting of an engine and van only, left Carnforth at 5.40 am, and at Grange coupled a rake of coaches. After a 20-minute halt, it continued as an all-stations to Barrow, for the benefit of dock workers. The last train from Carnforth was the 10.10 pm with through coaches off the 4.50 pm from Euston. It stopped at Grange and Ulverston only. This last train bore the nickname of 'The Whip', for its speed in passing through some of the stations. The up trains were similar, calling at all stations between Barrow and Carnforth, with the exception of the 9.15 pm which had through carriages for Euston, including a LNWR sleeping car, and which stopped en route at Ulverston and Grange only. In addition, there were five locals each way between Barrow and Ulverston, two of which continued to Grange, and three locals between Barrow and Millom, one of which started from Island Road station, by the shipyards, for the benefit of dock workmen. There was also a daily all-stations train from Millom to Whitehaven, and a market special between Barrow and Whitehaven on Thursdays only.

The Sunday timetable showed two trains, stopping at all stations, between Carnforth and Whitehaven, at 7.25 am and 5.20 pm, with corresponding return services at 10.10 am and 5.43 pm Only the 10.10 started at Bransty, all others starting or arriving at Corkickle; the reason for this was that the tunnel was closed from 10.15 am on Sundays for maintenance purposes. The remaining Sunday service consisted of two trains in each direction between Barrow and Carnforth, one evening trip from Barrow to Millom, and five in each direction between Barrow and Ulverston. During the summer only, three of the latter ran through to Lakeside. There was also one through trip from Barrow to Coniston, from June to September.

There were seven through trains on weekdays between Barrow and Lancaster, or Morecambe; and from June to September two from Lancaster to Lakeside, which travelled over the Leven Curve, avoiding Ulverston. These ran in connection with steamer sailings to Bowness and Ambleside.

Branch line services were quite diversified. On the Kendal branch on weekdays there were passenger trains from Grange to Kendal at 8.15, 10.00 am, 3.15 and 5.10 pm, with corresponding return workings at 8.50, 10.30 am, 1.20, 4.0, and 5.45 pm The 1.20 pm actually started from Oxenholme. The Kendal line also saw a fast goods for the North Eastern line at 6.05 am passing Arnside (4.40 am from Barrow). There were also the 7.55 am from Lindal Ore Sidings, 8.45 am and 10.15 am from Barrow

Yard; an empty train left Arnside at 12.38 pm for Oxenholme, and further fast goods were 1.30 pm from Barrow, 4.45 pm from Lindal, and 7.20 pm also from Lindal.

On the Windermere branch, there were passenger trains to Lakeside from Ulverston at 9.10 am, 10.40 am, 1.15 pm, 2.35 pm, 4.00 pm, 5.25 pm, and 6.35 pm In addition there were trains from Ulverston to Greenodd only at 7.05 am and 5.38 pm, and a mixed train at 8.20 am which ran non-stop to Lakeside. A goods train for the branch left Ulverston at 1.50 pm In the return direction, passenger trains ran from Lakeside to Ulverston at 8.30 am, 9.45 am, 12.30 pm, 1.55 pm, 3.25 pm, 4.35 pm, and 6.00 pm From Greenodd to Ulverston there was a short run at 6.30 am, and an empty train from Lakeside at 9.00 am; the Ulverston goods left Lakeside at 11.35 am

The Piel branch was served from Barrow with passenger trains at 7.55 am, 5.55 pm, and 8.45 pm, with an additional mixed train at 12.25 pm The return service ran at 8.13 am, 6.13 pm, and 9.03 pm, with a mixed train at 1.18 pm

The Coniston branch was well served from Foxfield, passenger train departures being at 8.55 am, 12.0 noon, 1.00 pm, 3.45 pm, 5.45 pm, and 7.00 pm, with additionally an empty train at 6.30 am and a goods at 12.5 pm In the opposite direction passenger trains ran at 6.00 am, 7.15 am, 10.55 am, 2.40 pm, 4.20 pm, and 6.15 pm, with goods trains from Broughton to Barrow at 8.30 am, and through goods from Coniston at 10.15 am

As far as goods services on the main line are concerned, there were eleven scheduled trains from Carnforth each weekday, all but two being designated semi-fast. The odd two were services around Lindal and a special goods train serving Kendal. One express goods to Whitehaven was for merchandise only, this train running over the original line, and thus avoiding Barrow. With only one stop at Millom, for water, it reached Whitehaven in three hours. It was usual to convey Millom traffic on Barrow trains as far as Lindal sidings, whence the wagons were worked forward by a local trip. Arrivals at Carnforth numbered twelve, of which one originated at Whitehaven, five at Barrow, four at Ulverston, one from Kendal, and one from Tebay. The up goods from Whitehaven was again express, but carried ore and coke for the Millom furnaces, stopping there to detach and pick up, the operation having a scheduled 35 minutes; this, however, was the only stop.

Coke traffic from South Durham was picked up at Tebay from the North Eastern Railway and worked forward by either Furness or LNWR locomotives.

There were two trips from Barrow to Whitehaven, both semi-fast, and a local pick-up from Barrow to Millom. Apart from these services, during the morning an engine and van left Whitehaven for Sellafield, where a train was made up from the WC&ER line and which then worked forward to Millom. By 1922 the mineral traffic from West Cumberland had fallen off considerably, and there was only one through train from Egremont to Millom, and vice versa. Remaining goods trips were purely local in nature, and included short workings between Millom and Lindal, Ulverston and the North Lonsdale Ironworks.

On both sides of Lindal, the gradients were severe, and loadings were restricted. From Plumpton Junction to Lindal summit there were almost four miles varying between 1 in 76 and 1 in 186; thence to Park South Junction, near Askam, a similar length of line fell at 1 in 74 and 1 in 144. On the climb from Barrow, Roose to Dalton rose steadily at 1 in 63. Banking engines were usually attached at Plumpton or Park South, or at Roose on the Barrow line. For many years the special heavy 0-6-0 tanks of Sharp, Stewart's design, affectionately known as 'Neddies', performed these duties until better replacements were afforded by Pettigrew's 0-6-0 and 0-6-2 tanks. The regulations limited the loads to 60 wagons on any train, including brake vans.

On the joint lines worked by the Furness Railway, and which comprised the former WC&ER from Whitehaven to Marron, and Moor Row to Sellafield, the Gilgarron branch, and other short mineral lines, the

A 'Neddie' No. 69 built in 1872.   *John Alsop collection*

regulations differed from those of the main line. The main joint line, the WC&ER, branched away at Corkickle, running parallel to the Furness Railway at first, then bearing away to the east at Mirehouse Junction, where there was a crossover between the two lines. Then came a stiff climb to Moor Row, the Corkickle Bank, with a ruling gradient of 1 in 52. Thereafter the double track continued to Rowrah and then became single to Marron. The Sellafield line started from Moor Row as single track to Woodend, then became double to Egremont, whence the remainder was single again. On this branch there were two sections of 1 in 80, falling towards Sellafield. The Marron line had gradients of 1 in 44 and 1 in 60 rising towards Rowrah, then falling again at 1 in 55 to Lamplugh.

Most stations on the joint lines were rebuilt after 1878, those on the Sellafield line to Furness design and the rest to LNWR standards. Signals and signal boxes throughout were of LNWR pattern. On single-line sections, the LNWR electric staff system was in use.

The bulk of the passenger traffic was worked by the LNWR, old four- and six-wheeled coaches being relegated to West Cumberland after living out their useful life on lines farther south. Beyond Rowrah, the LNWR handled all traffic, both passenger and goods. Three passenger trains in each direction worked on the Marron and Whitehaven section, supplemented by two return-trip goods trains. Latterly – in 1921 – the station at Marron was closed, and trains worked through to either

Rowrah station on the The Whitehaven, Cleator & Egremont Railway.

John Alsop collection

Workington Central or Workington Main. The Furness Railway also operated two passenger trains daily to Rowrah, two to Sellafield, and some special workmen's services between Beckermet Mines and Yeathouse. As with the LNWR, the Furness relegated its old four-wheeled coaches to these services. The Furness was also responsible for all mineral trips which did not run beyond Rowrah. On the Sellafield branch there was a large iron-ore and limestone traffic from the Beckermet and Biggrigg mineral branches, which was worked either to Whitehaven or Workington by the LNWR, or to Millom and Barrow by the Furness.

All the Furness locomotives working on the joint lines were housed at Moor Row shed. There were one or two 0-6-0s, both rebuilt and unrebuilt, of Sharp Stewart design, together with various 0-6-2 tanks specially built for the district, and two small 2-4-2 tanks for passenger traffic. The LNWR motive power consisted of five classes: 2-4-2 and "Coal" 0-6-2 tanks, 2-4-0 "Jumbos", and 0-6-0 tender engines of the "Cauliflower" and "Coal" classes. Banking duties on the joint lines were considerable, and for these the rebuilt WC&ER saddle-tanks were mainly used. Owing to the workmen's trains run over the Beckermet Mines branch, the electric staff had to be installed, between Mines Junction and Beckermet No. 1 Mine, in 1922. Part of the Gilgarron branch, between Ullock and Distington, was closed in 1917 after Wythmoor Colliery had been worked out. At the other end of the branch, a spur went off just before Parton to the Lowca Colliery and coke ovens. Originally the line from Parton to Distington was used for the conveyance of pig iron to Whitehaven for shipment, but this traffic practically ceased in 1918, and the only regular booked turns were between Parton and Lowca Colliery, these being worked by the LNWR.

To complete this survey of the joint lines it is necessary to include the section from Carnforth to Wennington. This was worked exclusively by the Midland Railway, the Furness never being responsible for more than the occasional ballast train for maintenance purposes. The Midland had its own shed at Carnforth, and ran passenger trains into a special bay at Carnforth station. Boat trains for Barrow and goods services ran direct over the line, avoiding Carnforth station, to the Exchange station at East Junction, where they were handed over to Furness motive power. Apart from the local passenger trains and the boat expresses, there were goods trains to and from the West Riding. There was one tunnel, at Melling, 1,230 yards long.

After the amalgamations in 1923, until 1927 the picture in Furness remained very much the same. Various older locomotives were scrapped and replacements were drawn from other directions or new engines built. On the branch lines, Lancashire & Yorkshire 2-4-2 tanks took over most services, whilst LNWR engines worked the main lines. Then came the Great Depression, which affected the whole country and which led eventually to the whole of the Furness and West Cumberland districts being designated a distress area. Under such circumstances, the London, Midland & Scottish Railway could do little but economise and retrench. So drastic was the loss of trade that a number of locomotives and rolling stock items became redundant and were withdrawn. Bus competition also became serious, with the result that all passenger services on the Cleator & Workington line, the joint lines between Moor Row and Marron Junction, and the Piel branch were withdrawn. The Lakeside branch was closed to passenger traffic in 1938, but was later reinstated, during the summer only. Many lines, such as the Gilgarron branch (except for the short section between Parton and Lowca Colliery) and "Baird's Line", as also the northern extension of the Cleator & Workington Railway through Calva and Great Broughton, have been dismantled. The direct line into the Barrow Haematite Steelworks was also severed by the removal of the junction at Hindpool, thus making it necessary to route all traffic in and out of the works via Barrow Central station and St. Luke's Junction. Since Nationalisation in 1948, further closures and dismantling of lines belonging to the old Furness Railway have been made, leaving little more than the main line through from Carnforth to Barrow and Whitehaven.

LMS camping coaches, old Midland and LNWR stock at Nethertown.

John Alsop collection

Chapter Seven

The Cleator & Workington Railway

As has been mentioned earlier in this book, when the WC&E Railway decided to increase their charges for minerals and coke, and were followed in quick succession by the LNWR and FR, the local ironmasters and merchants of West Cumberland were incensed. A meeting was called in Whitehaven early in 1874. The LNWR's name was mud in West Cumberland, and the WC&ER was losing its popularity. Lord Lonsdale was the prime mover of this new project, which was to build an independent railway to cut across the coalfield in direct opposition to the existing lines and provide a more direct route. Lord Lonsdale was backed by Baron Leconfield and Mr H.F. Curwen, of Workington, and these three whipped up considerable local support for their new line. The outcome was that a Bill for the construction of the Cleator & Workington Railway was put before Parliament in March 1875, and in spite of a lengthy hearing, with much bitter opposition from both the LNWR and Furness Railways, the Bill was passed. It might be of interest to quote from the contemporary local Press a comment on the "Big Bad Wolf" (the LNWR) which went as follows:

> The history of the LNWR in the district proves how extremely undesirable it is that the public should be placed at the mercy of a body of men whose only aim appears to be to get as much as they can, and give as little as possible in return. No sooner did the company get possession of the Whitehaven Junction and the Cockermouth & Workington Railways than the rates for season tickets were 'revised' which at Euston usually means 'increased'. In a short time there was another 'revision' of season ticket rates. The North Eastern Railway had a drivers' strike for a pay increase; there was none on the LNWR but the N.E.R. strike was taken as a pretext for rate increases on return tickets.
> In 1872, when the iron and coal trades were at their peak, carrying rates were nearly doubled. Now trade is depressed, but the high rates remain. As for the travelling public, they have been humbugged to all intents and purposes, and rickety carriages, condemned for the main line, are considered good enough for West Cumberland.
> Unpunctuality, defective management, high fares, exorbitant rates for traffic, scarcity of rolling stock, and second class passengers treated as so much rubbish-these are the leading features of LNWR policy in West Cumberland. But the North Western is not the only sinner, in respect of excessive rates. Both the Whitehaven, Cleator & Egremont and the Furness Railways are as deep in the mud as in the mire. The West Cumberland Blast Furnaces have to pay £9,000 more than a similar company in South Wales for hauling the same traffic over the same distance.

Of course the new line will be opposed tooth and nail, but Parliament will judge the Bill on its merits.

Much of this was undoubtedly true, but perhaps not as black as it was painted. Still, the local Press was backing the new company, and apparently thought it should support the venture as firmly as it could. Whether the Press actually had any influence on the issue is a moot point, but at any rate the Bill was "judged on its merits" as the paper said.

At a meeting at Workington in October 1875, it was announced that the route of the C&WR would be from Cleator Moor via Keekle to Moss Bay Ironworks and Workington, thence to Siddick, where it would join the LNWR line to Maryport. The contract for the construction of the line was let to Messrs. Ward & Co., and work on the new project commenced at Cleator. By an Act dated 28th June, 1877, the Furness Railway was empowered to work the C&WR line, and also included in the Bill were powers to offer to the Furness Railway the first option to purchase the undertaking should the present company wish to dispose of it at some future date. A further Act of 21st July, 1879, empowered the FR to take up shares in the C&WR. At first sight this offering of almost a free hand in the C&WR to the Furness Railway seemed a complete negation of the original purpose of the line, and was made even more curious by the fact

The Keekle viaduct under construction on the Workington and Cleator Railways main line which opened on 1879. *John Alsop collection*

that the Earl of Lonsdale, whose "baby" the C&WR undoubtedly was, also had a considerable influence in the WC&ER, which he sponsored some 20 years earlier, yet the offer was made to the FR, with whom the Earl had no connection at all and, indeed, had on occasions been at loggerheads. One can only assume that in the eyes of the more intelligent people of West Cumberland the Furness was the best of a bad lot, and if the line had to be leased to somebody, the Furness was the least of three evils.

A branch line was authorised in July 1878 from Distington to Rowrah, making the third side (though by no means direct) of the WC&ER triangle: Distington-Ullock-Rowrah; and a short branch in Distington Ironworks was put in early in 1881. The main line from Cleator Moor to Workington and Siddick was opened on 18th October, 1879.

In 1883 powers were obtained to construct what was termed the Northern Extension, from Workington to Brayton (15¾ miles) on the Maryport & Carlisle main line, but this was only constructed as far as Linefoot, where it joined the M&CR Bullgill-Brigham branch. Powers to abandon the remainder of the line to Brayton were granted in 1886. Ultimately the last two miles of the Northern Extension, from Buckhill Colliery to Linefoot, were also abandoned in 1915, and the line beyond Buckhill dismantled. In its final form, the C&WR consisted of the main line from Cleator Moor to Siddick (11½ miles), the Distington-Rowrah

Workington Central station surrounded by bunting for Queen Alexandra Floral Day in June when a collection was made for charity. Today the Alexandra Rose Charity's work continues raising money to help with food poverty. *Oakwood Press*

The Moss Bay Ironworks connected to the C&WR through its mineral lines. It was established by Charles James Valentine who was also a director of the railway.
Oakwood Press

branch (6½ miles), and the Northern Extension to Buckhill (4 miles). In addition there were four short mineral lines, to Harrington Ironworks (2 ¾ miles), the Moss Bay and Derwent Ironworks (approximately 1 mile each), and 1½ miles of the Lowca mineral branch from Harrington to Rosehill Junction, from which latter point to Lowca Colliery was the property of the Workington Iron & Steel Co.

The whole system was single line, operated by the electric staff of LNWR standards done to conform with single-line operation throughout the area, not from any love for the LNWR. Exceptions to this were the Moss Bay and Derwent branches, which were worked by staff and ticket.

Situated as it was on the fringe of the mountainous Lake District, it is not surprising that gradients were severe. From Cleator Moor, after a short length of 1 in 284, there was almost a mile of 1 in 72 and 1½ miles of 1 in 70 rising to Moresby Parks. After a short level section, a falling gradient of almost three miles of 1 in 70 brought the line to Distington and, with short breaks, this gradient continued down to Workington, a further three miles. There were intermediate stations at Cleator Moor (C&WR), Moresby Parks, Distington, and Harrington. There was also an unstaffed halt at Keekle, for workmen at Moresby Colliery. On the Northern Extension there was a continuous rise of two miles from the junction (Calva), again at 1 in 70. The only passenger station on this line was at Seaton. However, the most severe gradients occurred on the cross-country Distington-Rowrah line. This branch was built primarily to connect with a privately owned mineral line, owned by the Scottish firm of Bairds Ltd., and named the Rowrah & Kelton Fell Mineral Railway. Locally it was always known as "Bairds' Line", and in due course this cognomen became attached to the whole branch from Distington. Except for the last short piece from Rowrah to Arlecdon, the whole of "Bairds' line" was dismantled in 1939. From Distington the line began with a short rise of 1 in 70, then two miles of 1 in 44 to Oatlands, where there was a small station and a colliery, then after a short length of almost level, there was a mile of 1 in 52 to the summit, at about 600 feet above sea level. From here it dropped at 1 in 60, crossing the bottom of a valley on a high embankment on a 14 chain curve, then climbed up the opposite side at 1 in 64 for almost a mile before the final drop at 1 in 72 to Arlecdon and Rowrah. The whole branch abounded in sharp curves, and coupled with the gradients made it extremely difficult to work.

Passenger traffic was almost negligible and came to about five short trains each way between Cleator and Siddick. On market days and Saturdays there was an extra service between Workington (C&WR) and

The modest station buildings at High Harrington.   *Oakwood Press*

Seaton, on the Northern Extension, and between Distington and Rowrah. This latter service consisted of one through train from Rowrah to Workington at midday, returning in the evening. Miners' trains were run as necessary over most of the system, including the mineral branches. While iron and coal trades prospered, the line carried a heavy traffic. All coke for the blast furnaces at Cleator and Distington was brought from South Durham via Siddick, and pig iron was taken out by the same route. Coal traffic from the collieries at Moresby, Broughton Moor, Buckhill, Harrington, and Lowca was also heavy, while limestone and iron-ore were brought from Lamplugh and Rowrah.

During the 1914-18 war, some unusual methods of train working were in force to cope with the extra heavy traffic. It was quite common to see two trains (or even three) coupled together, and with one or more banking engines in the rear, climbing up from Distington to Moresby. Thus one was frequently treated to the spectacle of a long train with engines at front and rear, and with one or more engines interspersed among the wagons as well. There appear to have been no set restrictions on the length or weight of trains, as long as adequate engines were available. There were, however, very severe restrictions about descending trains, in order to guard against the possibility of breakaways; the 18" Furness 0-6-0 goods engines, for example, were limited to 45 wagons loaded, or 60 empty, on gradients exceeding 1 in 70.

Dissington station was situated at the junction between the Cleator and Workington Railway's main line (right) and the Furness/LNWR joint line from Whitehaven.

John Alsop collection

THE CLEATOR & WORKINGTON RAILWAY

There were no Sunday trains, except for a couple of miners' services on the Lowca and Derwent branches, these being worked by the Lowca Coal Co.'s own locomotives.

The C&WR had ten engines of its own, all saddle-tanks, though there were never more than eight in stock at any one time, and these worked mostly at the Workington end of the line. The Furness provided all other power from their shed at Moor Row, though they also stabled a couple of engines at Workington. All coaching stock, mostly decrepit six-wheelers, and a large proportion of the goods stock was provided by the Furness. There were a number of C&WR 12- and 15-ton open wagons and some 10-ton wagons with dumb buffers, which were all painted red oxide, with C&WR in large white letters. The dividends paid by the company were remarkably consistent, never less than 3% and never more than 4½%, The option to purchase by the FR was never exercised, and the C&WR remained nominally an independent company until the Grouping.

Cleator & Workington Railway Locomotives

The first locomotives which worked on the line appear to have been the property of the contractors and were never taken into stock. These were two 0-6-0 saddle-tanks of Manning Wardle standard design, with 3 ft 6 in. wheels and inside cylinders 12 in. x 17 in. They were named *Derwent* and *Keekle*, makers' numbers 679 of 1877 and 684 of 1878. These were followed by two small 0-4-0 side-tanks, *Brigham Hill* and *Flosh*, which were local products, having been built by Fletcher Jennings & Co. at the Lowca Engine Works, their numbers 187 and 152, built in 1882 and 1876 respectively. *Flosh* was said to have been sold in 1878, but *Brigham Hill* remained in stock until 1897.

The stock proper can be said to have begun with No. 3, *South Lodge*, an 0-6-0 saddle-tank of Robert Stephenson & Co.'s standard design, which was delivered in 1884 (works No. 2553). In dimensions it was almost exactly like the WC&ER saddle-tanks of 20 years earlier, but rather more modern in appearance. Wheels were 4 ft 6 in. diameter, and inside cylinders 17 in. x 24 in., and the boiler worked at 140 lbs. The saddle-tank extended over the firebox but left the smokebox uncovered, and the cab was open backed. This engine worked until 1920, when it was withdrawn. A further odd engine followed in the next year, similar in general appearance, but differing slightly in dimensions, which were 4 ft 3 in. for the wheels and 16½ in. x 24 in. cylinders, with working

One of the earliest C&WR locomotives *Brigham Hill* was built in 1894 by Fletcher Jennings Ltd of Lowca. It was renamed *Rothersyke* in 1893 and sold to the West Stanley Colliery in Co. Durham in 1897.
John Alsop collection

Cleator & Workington Locomotive No. 3 *South Lodge*, a Robert Stephenson & Co. 0-6-0.
John Alsop collection

pressure 160 lbs. This engine, No. 4, *Harecroft*, was built by the Lowca Engine Co. (successors to Fletcher, Jennings) in 1885, their No. 196. It was handed over to the Moss Bay Iron & Steel Co. in 1913, it is often said "for debt", but in view of the good financial position of the C&WR throughout its career, this seems hardly likely. It seems more credible that there was a direct sale, by mutual arrangement.

Three further Stephenson 0-6-0 saddle-tanks were added to stock, all identical with No. 3, except their working pressure was 160 lbs. and that they had overall cabs. These were respectively 5, *Moresby Hall*, Stephenson 2692 of 1892, 6, *Brigham Hill*, R.S. 2813 of 1894, and 7, *Ponsonby Hall*, R.S. 2846 of 1896. (The original *Brigham Hill* had been renamed *Rothersyke* in 1893.) No further stock changes took place until 1897, when the company departed from their long association with Stephenson's and ordered a similar engine from Peckett & Sons, of Bristol. This was somewhat larger in dimensions, with 18 in. x 24 in. cylinders and 180 lbs. pressure, though the wheel diameter remained the same. This engine, and an exactly similar one ordered in 1917, had full-length saddle-tanks. The final addition came in 1920, from Hudswell Clarke, of Leeds, generally identical, but having 200lbs. pressure. In turn these engines were named and numbered 8, *Hutton Hall*, Peckett 1134 of 1907, 9, *Millgrove*, Peckett 1340 of 1917, and 10, *Skiddaw Lodge*, H.C. 1400 of 1920.

Nos. 6 to 10 came into LMS stock in 1923 and were renumbered 11564

The second *Brigham Hill* No. 6 was a sister engine to No. 3   *John Alsop collection*

to 11568. The only alteration their new owners made was to promptly reduce the working pressure of No. 10 to 160 lbs. No. 11564 was withdrawn in 1926, 11565/6 in 1927, 11567 in 1928. 11569 lasted until June 1932, when it was sold to the Lambton, Hetton & Joicey Collieries.

All engines were painted a middle green, with black bands edged with fine red lining. The name and number plates were polished brass with raised lettering, the names being fixed to the centre of the saddle-tanks and number plates on the bunker. No. 10 was delivered with "Cleator & Workington Railway" painted on the tanks, but Mr Murray, the locomotive superintendent, promptly had it painted out. All engines from No. 7 onwards were fitted with the automatic vacuum brake for working passenger turns, and all were fitted with a second, lower set of buffers for use with the small local colliery wagons.

The last C&WR locomotive, No. 10 *Skiddaw Lodge*, built by Hudswell Clarke.
John Alsop collection

Chapter Eight

Furness Railway Locomotives

Until 1896, the Furness Railway had no locomotives of its own design; all had been the standard productions of the various makers from whom they had been purchased. By far the largest number were built by Sharp, Stewart & Co., of Manchester, and in this the FR was closely analogous to the Cambrian Railways, both companies having a number of engines of the same design and dimensions. The Furness engines were of seven types, 0-4-0, 0-6-0, 2-4-0 and 4-4-0 tender, and 2-2-2, 0-4-0, and 0-6-0 tanks. The first four engines, however, were built by Bury, Curtis & Kennedy, of Liverpool, and were of the famous bar-framed 0-4-0 type one of these, No. 3, is still preserved, though not now in working order, since she suffered damage during one of the bombing raids during the last war, when her glass pavilion at Barrow Central station was shattered by the Luftwaffe. Removed afterwards to Horwich Works for safety, she stood in the paint shop there for several years, before finding a place in the British Railways Museum at Clapham, and later moved to York when the National Railway Museum opened in 1975.

From 1896, when Mr R. Mason retired from the post of locomotive superintendent, and was succeeded by Mr W. F. Pettigrew, the company

Furness Locomotive No. 3 'Coppernob' being removed from the glass pavilion at Barrow station in 1938 to travel to the London & Birmingham Centenary exhibition in Euston station. The locomotive also visited London for the British Empire Exhibition in 1925 to be a contrast to the latest engines. *John Alsop collection*

had its own designs, which were eminently suited to the work required of them, culminating in the large "Baltic" express tank engines of 1920. By then Mr Pettigrew had in turn retired, and the Baltics were the sole design produced by his successor, Mr D. L. Rutherford, before the Grouping put an end to the Furness identity.

There was no official classification of locomotives. There is, however, a classification in existence which has been more or less accepted, is believed to have been introduced by that eminent historian, the late A. C. W. Lowe. As this classification is very useful in sorting out the different types of locomotives from the very early days, it will be continued here, but it must be emphasised that it is not official.

A severe complication in the Furness stock list is the great amount of renumbering which took place at various intervals, mainly to maintain similar locomotives in the same numerical sequence. There was, too, a spasmodic use of an "A" list for engines replaced but still retained in service. All this has needed a great deal of unravelling, but the version given is believed to be as near correct as it is possible to be.

It is proposed to deal with the locomotives in type order, not chronologically, as it was thought that by so doing the evolution of each type will be easier to follow. Details of some of the former Whitehaven & Furness Junction engines taken over in 1866 are very elusive, and even the wheel arrangement of some has been in dispute. The version of the list in the present publication is that of the late A. C. W. Lowe, modified in the light of some recent researches by Mr E. Craven into the early locomotives of the Lancashire & Yorkshire Railway, for it has been established by Mr Craven that certain L&YR engines were sold to the W&FJR. Even so, the absolute accuracy of this list cannot be guaranteed. This book is not really concerned with the W&FJR locomotives which passed to the LNWR in the division of stock in 1866, but they are included for the sake of completeness. The LNWR numbers allotted (though several of them never actually carried) have been the subject of no little controversy, and no two authorities seem to agree. The first series allotted was 1551-1560, altered within a few weeks to 1578-1587, though in what order is problematical.

Class A1. 0-4-0 tender engines

There were two engines of this class, numbered 1 and 2, which were delivered in 1844, and used in the construction of the line. Built by Bury, Curtis & Kennedy, they had the typical bar frames and hemispherical "haystack" fireboxes, with domeless boilers, for which Bury's engines were famous. In fact, except for a small number of engines built to customers' own designs, all the firm's output of some 350 locomotives until they ceased production in 1850 were of this same basic design. Nos. 1 and 2 were smaller engines than the well-known 'Coppernob' of two years later, and had a much shorter life. No. 1 had her firebox badly burned at Carnforth in 1866 through the fire having been lighted with an empty boiler, and she was broken up shortly afterwards. Her sister was sold to a Northumberland colliery, and her subsequent history is unrecorded. The internal dimensions of these two engines were open to doubt. Their cylinders, inside, and hung low, were 13 in. x 24 in., wheels 4 ft 9 in., wheelbase 7 ft 5 in., and working pressure 90 lbs. The four-wheeled tender, of typical Bury pattern, carried 900 gallons of water and 2 tons of coal.

Class A2. 0-4-0 tender engines

This class also comprised two engines, Nos. 3 and 4, by the same builders, but delivered in 1846, and were rather larger than Class A1. These four engines sufficed for some six years and worked all the traffic quite successfully over the two short lines which made up the Furness Railway at that period. Frames were the usual bar pattern; the boiler was of Low Moor iron and was domeless, working at 110 lbs. The firebox was of copper, with the upper part hemispherical, and was surmounted by a small dome carrying two spring balance safety valves. Originally

A2 locomotive No. 4, built by Bury, Curtis & Kennedy of Liverpool in 1846 and withdrawn in 1898.   *John Alsop collection*

the firebox crown plate was made in two halves, lap-jointed, but this proved somewhat troublesome and was replaced in 1853 by a single plate, which gave no more trouble. Cylinders were between the frames and were cast separately, the Stephenson link motion working above the leading axle. The tender had oak frames, with axle-guards bolted on, and ran on four wheels of 3 ft 1½ in. diameter, at 6 ft 8 in. centres. Water capacity was 1,000 gallons and coal 1½ tons.

Latterly, the two engines were employed in shunting Barrow Docks and on short local goods trips. No. 4 was taken out of service in 1898 and broken up soon afterwards, but No. 3, withdrawn in 1900, was at first stored in Barrow Works, but in 1907 was mounted on a pedestal at the Central station.

Class A3. 0-4-0 tender engines

As the Bury four-coupled engines had proved quite capable of dealing with the mineral and goods traffic, when further engines were required four more of the same general pattern were ordered from W. Fairbairn, of Manchester. Bury's had gone out of business by this time, but

A3 locomotive No. 9 delivered to the Furness Railway in 1855 and the last of the class when withdrawn from service in 1901. *John Alsop collcection*

Fairbairn had built a number of Bury engines under sub-contract, and therefore had all the necessary patterns and specifications. Two were delivered in 1854, numbered 7 and 8, while numbers 9 and 10 followed in 1855. While of the same general design as the two previous classes, they were somewhat larger in dimensions, the chief external difference being that they had closed splashers over the wheels. All four engines put in more than 40 years' work, Nos. 7 and 8 being withdrawn in 1899, No. 10 in 1900, and 9 in 1901. The last two were renumbered into the duplicate list as 9A and 10A in 1899. Their tenders were of Fairbairn's own design, with a peculiar pattern of plate frame, in which the support for the brake gear was carried in a curved part of the main frame, centrally between the axles. The lower part of the frame was thus made up of two semi-circles. Springs were below the running plate, and the body of the tender was based on Bury's design. The tender ran on four wheels at 9 ft centres, and carried 1,000 gallons of water and 2 tons of coal. The engines had larger cylinders than the Burys 15 in. x 24 in. and the wheels were 4 ft 6 in. diameter.

Class A4. 0-4-0 tender engines

Four further engines were ordered from Fairbairn and were similar to the previous class. The splashers were of a different pattern. Nos. 13 and 14 were delivered in 1858, followed by 15 and 16 in 1861. Dimensions

A4 locomotive No. 16. *John Alsop collection*

were the same as Class A3, except that the weight was 24¼ tons. Nos. 15 and 16 were taken out of stock in 1899, but 13 and 14 were put on the A list in the same year and were withdrawn in 1900. All the 0-4-0s were replaced in the stock list by Pettigrew 0-6-0 tender engines.

Class A5. 0-4-0 tender engines

The last four-wheeled tender engines were provided by Sharp, Stewart & Co., since Fairbairn's had, in their turn, gone out of business, and their stock and goodwill were purchased by Sharp, Stewart. Nos. 17-20 were built in 1863, Nos. 25 and 26 in 1865, and in 1866, Nos. 27 and 28. No further 0-4-0 tender engines were purchased. The eight members of this class were of the builders' own design, considerably different from the earlier 0-4-0s. Plate frames were used; the boiler had a dome on the front ring, surmounted by Salter safety valves, and the firebox was the normal round-topped pattern with raised casing. Cylinders were 15 in. x 24 in., with valves on top, actuated by Stephenson link motion. The footplating was raised over the wheels, with a very deep valance. The engines were fitted with standard Sharp four-wheeled tenders, of which the Furness Railway eventually had a considerable number, since a modified version was also fitted to the later 2-4-0 and 0-6-0 engines built by the same firm. The springs were above the running plate, which gave a rather narrow body, and there was a large toolbox on the rear framing above the buffer beam.

A5 locomotive No. 27. *John Alsop collection*

Six engines of this class were sold in 1870 and 1873 to the associated Barrow Haematite Steel Co., thus having a comparatively short career with the FR This was not because they were not good engines; far from it, for they gave good service in the steelworks for many years, in fact some reached the ripe old age of 80 years. They were altered by Sharp's to saddle-tanks before being handed over. Furness trains were rapidly becoming too heavy for these small engines, and their work could be performed better by the 0-6-0 type which were just coming into their own. For such light work as was within the capacity of the four-wheeled engines, the older Bury and Fairbairn classes were adequate, and it was considered better to sell the Sharp engines while they were comparatively new and could command a good price. Nos. 27 and 28 were retained by the railway, and after being renumbered 27A and 28A in 1914, were finally taken out of service in 1918. Sharp's works numbers were 1434/5, 1447/8 of 1863 (Nos. 17-20); 1585/6 of 1865 (Nos. 25/6); and 1663/4 of 1866 (Nos. 27/8).

Class B1. 2-2-2 well-tank engines

Up to 1850, the Furness Railway had never bothered much about passenger. Only four passenger coaches had been ordered when the line was opened-but with the possible extension to Ulverston and beyond, more notice had to be taken of potential services. So two small 2-2-2 well-tank engines were ordered from Sharp Brothers in 1851. They were delivered in the following year, numbered 5 and 6 in the Furness list, and 696/7 in the makers' books. They were very small locomotives, of

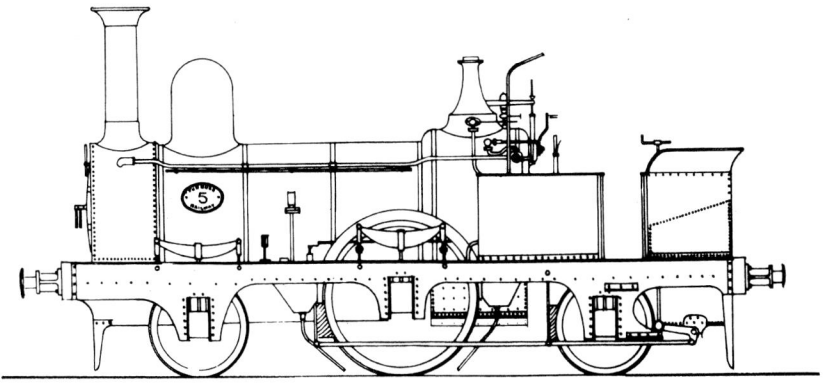

B1 locomotive No. 5   *Oakwood Press*

Sharp's standard design, with outside sandwich frames, inside cylinders 14 in. x 18 in., and a small water tank under the bunker. The framing of this pair was straight, unlike the subsequent engines of the same type which had the framing curved over the driving axle. Driving wheels were 5ft. 6 in., and the carrying wheels 3 ft 6 in. No cab was fitted, but later a bent-over iron sheet, serving as both roof and spectacle plates, was fitted. The boiler was the usual Sharp pattern, with dome on the front ring and Salter safety valves on the raised firebox. Both engines were withdrawn in 1873.

Class B2. 2-2-2 well-tank engines

Two further small passenger engines were obtained from Sharp Bros. in 1857, their Nos. 1016/7, becoming 11 and 12 on the FR list. They were similar in most respects to the earlier pair, but differed in having the outside frames curved over the driving axle, and were slightly larger in general dimensions. Both had cabs of a sort, a single sheet of metal bent over to cover the footplate and strengthened by angle-iron. No. 11 was sold back to Sharp, Stewart in 1873 in part payment for a new engine, but No. 12 lasted considerably longer, being renumbered 12A in 1873, and was finally sold to the Weston, Clevedon & Portishead Railway in 1898. This engine worked for some time on the Coniston branch and was

B3 locomotive No. 37.   *John Alsop collection*

involved in a collision between Broughton and Foxfield while on this duty; this incident is mentioned later, in the chapter on accidents.

Class B3. 2-2-2 well-tank engines

This was the largest group, both numerically and in dimensions, of all the Furness engines of this type. There were six, Nos. 21 and 22 built in 1864, and 34-37 built in 1866, all by Sharp, Stewart & Co. The works numbers were 1500/1 and 1763/68/67/62. They were exactly like Class B2 in appearance but somewhat larger in dimensions, proving versatile and economical for light passenger duties. They had the same wheel diameters as the earlier engines, but their cylinders were enlarged to 15 in. x 18 in., and they had larger boilers. All had a fairly long career; they were put on the A list in 1896 and withdrawn two years later. No. 35A was sold to the Weston, Clevedon & Portishead Railway, and 22A to the South Shields, Marsden & Whitburn Colliery Railway; Nos. 36/37 worked as stationary engines until 1918; the rest were scrapped.

Class B4. 2-2-2 well-tank engine

This was an odd engine, No. 10, *Queen Mab*, of the Whitehaven & Furness Junction Railway, built by R. & W. Hawthorn (No. 1128) in 1860. It became No. 46 in the Furness list and did not last long, being sold to the Isle of Wight Central Railway in 1876, where it was renamed Newport, and ran until 1890. The engine was similar in dimensions to

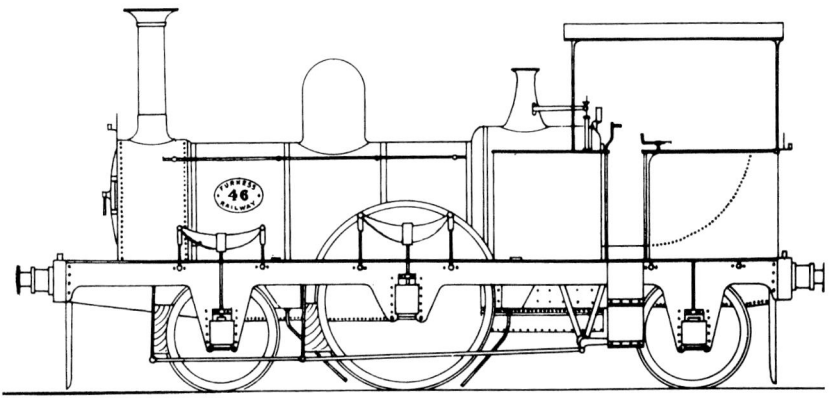

B4 locomotive No. 46. *Oakwood Press*

the Sharp engines but differed considerably in appearance. The plate frames were outside, and were straight; the boiler was domed and the safety valves were in a brass casing over the firebox. There was an all-over cab, without side-sheets, which embraced the bunker, and water was carried in tanks fitted between the main frames. For some time after the amalgamation, No. 46 remained on the Whitehaven section but was then transferred to Barrow, and worked for the rest of its career on the Coniston and Lakeside branches.

Class B5. 2-2-2 well-tank engines

These two engines were part of the W&FJR stock, in which they were No. 4, *Oberon*, and 5, *Titania*, being renumbered into the Furness list as 47 and 48. Both were built by E. B. Wilson & Co., of Leeds, in 1850. They were small well-tank engines, typical of the makers' design, and unlike the other 2-2-2 tanks had inside frames. Driving wheels were 5 ft 3 in. diameter, and the inside cylinders were 12 in. x 18 in. Quite capable of pottering about on the almost level Whitehaven line, they were not so useful on the Furness Railway proper, and spent what little time they were on the FR books doing odd jobs on the smaller branch lines. Both engines were sold in 1870. Incidentally, all engines of the W&FJR. carried names as well as numbers, but when the Furness Railway took over all names were removed. The same applied to the locomotives of the WC&ER. No Furness engine officially carried a name.

B5 locomotive ex-Whitehaven & Furness Junction Railway No. 4 *Oberon*. *Oakwood Press*

C1 locomotive No. 23. The dome over the firebox nestled in the cab. *John Alsop collection*

Class C1. 0-4-0 saddle-tank engines

For shunting in the docks area and work on mineral branches, Sharp, Stewart & Co. supplied two four-wheeled saddle-tank engines of their standard design in 1864. Numbered 23 and 24, they were 1543/4 in the makers' list. They had inside cylinders 14 in. x 20 in., inside frames, and 4 ft wheels. The boilers were domeless, but a large dome surmounted by a pair of Salter valves was fitted over the raised firebox. The saddle-tank covered the smokebox but stopped short at the firebox throat plate, and the usual primitive bent sheet of metal served for a cab. Four further engines of the same class, but having the dome on the first ring of the boiler instead of over the firebox, were delivered by Sharp's in 1874, numbered 94-97 in the FR list and carrying Sharp's works Nos. 2448-51. In 1898, 23 and 24 were renumbered 98 and 99 to make way for a batch of Pettigrew 0-6-0 goods engines, but did not last long after this, being withdrawn in 1904. One of the later batch, 97, actually lasted long enough to be taken over by the LMS, who scrapped it in 1924, under the number 11258 which it never actually carried. At some time in its declining years, this engine acquired a pair of Ross pop valves on the dome, in place of the original Salter type, as did 95. Locomotives 94 and 95 were put on the A list in 1912, the former being scrapped in 1914 and 95(A) two years later. No. 96 was on the A list in 1907 and was also scrapped in 1916.

Class C2. 0-4-0 saddle-tank engines

About these two engines very little is known. Though classed together, they had little in common and came from different makers. Both were built for the Whitehaven & Furness Junction Railway, No. 15, *Banshee*, by Fletcher Jennings & Co. (works No. 29) in 1862, and No. 16, *Bob Ridley*, by Neilson & Co. (works No. 571) in the same year. The Furness Railway renumbered them 49 and 50. Apart from the fact that they had 10 in. x 16 in. cylinders and 4 ft wheels, nothing has survived of their dimensions. Used for the transfer service through the streets of Whitehaven, between Preston Street goods station and the harbour, they were withdrawn from service in 1882, and were both sold to a local Whitehaven firm of contractors.

Class D1. 0-6-0 tender engines

This was the largest class, numerically, of all the Furness locomotives, totalling no less than 55, plus eight further examples of the class which were ordered by the company but cancelled before delivery. In addition, this class was the most complicated, owing to subsequent

D1 locomotive No. 53 with the original boiler design. *John Alsop collection*

FURNESS RAILWAY LOCOMOTIVES 111

D1 locomotive No. 40 rebuilt in 1900 to the second boiler design. *John Alsop collection*

rebuildings. The increasing weight of mineral trains was getting beyond the capacity of the Bury and Fairbairn four-wheeled engines, and Sharp, Stewart & Co. were asked to submit designs for a larger type of engine. The type chosen was a medium sized 0-6-0 tender locomotive, of which the makers built a considerable number, both for home and overseas. The first example was built in 1866 and the last in 1884, during which period of almost 20 years no modifications were made to the design and little to the dimensions. They were small machines by modern standards but were very efficient and cheap to maintain and they were very popular with the crews. Their wheels were 4 ft 6½ in. diameter and inside cylinders 16 in. x 24 in. The boiler was of average size for the period, and had the usual Sharp raised firebox with Salter valves in a neat brass casing on top of it. The dome was on the middle ring of the boiler. All those built before 1873 had no cab, only a front weatherboard, but the later engines had rather a sketchy cab, similar to the Stirling pattern on the Great Northern Railway. The tenders were four-wheeled, of the same design as those of the contemporary 2-4-0s, holding 1,500 gallons of water and 3 tons of coal.

When W. F. Pettigrew took charge of the motive power in 1896, he commenced bringing them up to date, fitting automatic vacuum brakes and steam heating for use on passenger trains, also fitting larger steel boilers of his own pattern, standard with the L1 class of 0-6-2 tanks but with a smaller firebox. These boilers had flush fireboxes and Ramsbottom safety valves. At the same time new cabs were fitted, of more orthodox type, but retaining the large oblong rear splasher, with cut-out side sheets and a separate roof plate. Still later, between 1910 and 1912, several were rebuilt with boilers standard with Pettigrew's 0-6-0 side-tanks, but without the extended smokeboxes. Finally, around 1916, five members of the class received similar boilers, with extended smokeboxes, and completely new cabs, in which the oblong rear splasher was discarded.

No engine was rebuilt more than once, and it is probable that those scrapped before 1915 retained their original boilers to the end. All engines which were taken over by the LMS in 1923 had been rebuilt, and those renumbered in the series 12000-12014 had the 1898 pattern boiler, while those in series 12065-12076 had the larger boilers with or without extended smokeboxes. The larger boilered engines did consistently good work all over the system, both on passenger and goods trains. However, none of them lasted very long under LMS rule; the whole class was extinct by 1930.

Of the ten engines ordered by the Furness Railway, but cancelled, two actually did run on FR metals; these were Nos. 120 and 121, delivered in 1884 and sold in 1887 to the Liverpool, Southport & Preston Junction Railway. They eventually passed to the Lancashire & Yorkshire Railway when that company took over the Liverpool, Southport & Preston Junction Railway in 1897, but were scrapped without receiving the L&YR numbers allotted (1371/2). Of the other eight, two (Sharp Stewart 2339 and 2347 of 1873) went to the Mid Wales Railway, eventually becoming Cambrian Railways 49 and 50 in 1888; two (Sharp 2342 and 2346) became North Staffordshire Railway 69 and 70; two went direct to the Cambrian Railways (Nos. 14 and 15, Sharp 2511 and 2513 of 1875); one to the Denbigh, Ruthin & Corwen Railway (Sharp 2510 of 1875), passing to the LNWR and finally to the Cambrian as No. 18; the last, Sharp 2512 of 1875, was sold by the makers to an unknown customer overseas.

It was one of this class, No. 115, in its original condition, which was lost in the Lindal subsidence of 1892, which unfortunate occurrence will be dealt with in the appendix on accidents. Several times proposals to recover the engine were brought up, but abandoned owing to the great depth to which it had sunk, estimated to be about 200 feet below track level. The tender broke away from the engine and did not fall into the hole, though it was derailed, and was recovered.

D1 locomotive No. 115, originally No. 114 but renumbered in 1898 after the first No. 115 was lost in the Lindal subsidence of 1892. The locomotive was rebuilt in 1910 to the third boiler design and new cab. *John Alsop collection*

D1 locomotive renumbered twice from 26 to 59 (1913) and eventually No. 63 (1918). The locomotive was rebuilt in 1916 with the fourth boiler design. *John Alsop collection*

D2 locomotive No. 42 after alterations to fit a dome. *Oakwood Press*

Class D2. 0-6-0 tender engines

The two engines of this class, Nos. 42 and 43, were acquired from the Whitehaven & Furness Junction Railway, on which line they had been numbered 19, *Lonsdale*, and 18, *Cedric*, respectively. Both were built by R. & W. Hawthorn in 1864, the works numbers being 1269 and 1245. Designed by Mr Meikle, the locomotive superintendent of the W&FJR, they had a rather- squashed appearance, owing to the unusually short front and rear overhang. This shortness in overall length had its effect on the boiler dimensions, which were only moderate, slightly less, in fact, than the Sharp engines of Class D1. As originally built, they had domeless boilers and raised fireboxes on which the dome was mounted. Frames and cylinders (16 in. x 24 in.) were inside; wheels were 4 ft 6 in. There was no cab, only a scanty front weatherboard. No. 42 had a longer life on the FR than any other W&FJR engine; it was rebuilt at Barrow in 1886 with a domed boiler, probably the old boiler retubed and fitted with a dome, and lasted until 1904. No. 43 was soon disposed of, being sold in 1873 to Fletcher, Jennings & Co., subsequently finding its way to the Wigan Coal & Iron Co., where it ended its days.

Class D3. 0-6-0 tender engines

Pettigrew soon saw that the small Sharp goods engines had insufficient power to cope with the ever-increasing weight of the iron-ore trains, though they could still do good work on the lighter general merchandise trains, and he proceeded to design a much more powerful engine to take their place. He was a great believer in standardisation, and so as many parts as possible were made interchangeable with the 0-6-2 tanks he had already produced for the Cleator Moor district. The boiler, cylinders, wheels, and motion therefore were the same for both classes. Twelve of these engines were ordered, the delivery being shared equally by Sharp, Stewart & Co. and Nasmyth, Wilson & Co. The running numbers allotted were 7-12 for the Nasmyth batch and 13-18 for the Sharps, works numbers being 552-57 and 4563-68 respectively, all dated 1899. In the Furness list they replaced a motley collection of old engines, mainly 0-4-0, some of which were put on the A list and others scrapped, but 11 and 12 were of the Sharp 2-4-0 passenger class, and were renumbered 3 and 4. All twelve of the new engines were put to work on the Tebay-Barrow coke trains, the Carnforth-Whitehaven express goods, and others of the heaviest trains, releasing the Sharp goods engines for secondary duties.

Two of the class were fitted with Macallen's variable blast pipe, but this fitting was removed after a short trial, as it did not prove to have any appreciable effect on the steaming. Automatic vacuum brakes and steam heating were fitted for use on passenger trains; though they were not used regularly for passenger services, they were pressed into use for

D3 locomotive No. 7 passing Ravenglass station with a goods train.

John Alsop collection

specials and excursion work on odd occasions, their 4 ft 7½ in. wheels being on the small side for fast running. Some of their hardest work was done in the Whitehaven area, on the "Joint Lines" trains. The design was simple and straightforward, with short smokebox, round-topped firebox, and dome in the centre of the boiler; the Ramsbottom safety valves were surrounded by a neat casing, which became the standard for all FR engines. Cylinders, 18 in. x 26 in., were inside and operated by Stephenson link motion. All the class were in service at the Grouping, and were given LMS numbers 12468-12479, in order of FR numbers. Two, 12469 and 12479, were rebuilt at Horwich with L&YR pattern saturated Belpaire boilers and extended smokeboxes. The first to be withdrawn was 12468 in 1928 and the last, 12479, in August 1936.

Class D4. 0-6-0 tender engines

Following the success of his first class of mineral engines, Pettigrew introduced a mixed traffic version, with 5 ft 1 in. coupled wheels. Only four of these were built, by the North British Locomotive Co. (as successors to Sharp, Stewart & Co.) in 1907. They were Nos. 3-6 in the Furness list, and carried NBL works Nos. 17840-3. Actually, Pettigrew introduced this class in 1900, and six engines were ordered from

D4 Locomotive No. 3. *John Alsop collection*

Nasmyth Wilson in that year, but owing to a temporary falling-off in traffic the order was cancelled. The six engines were completed by the makers (works Nos. 588-593) and were sold to the North Staffordshire Railway, becoming their 159-164.

In dimensions they were similar to the D3 class, but were also standard as far as possible with the 0-6-2 tanks of Class L2. Two were stationed at Whitehaven, and the other pair at Carnforth, for fast goods work and also for occasional use on passenger services, for which they were· fitted with vacuum brakes and steam heating apparatus.

Three, Nos. 3, 5, and 6, were rebuilt in 1926 with L&YR saturated Belpaire boilers and extended smokeboxes, but apart from No. 5, which was withdrawn in 1934, this did not extend their life, as the other three, including the unrebuilt No. 4, were all withdrawn in 1930. The LMS gave them Nos. 12480-83, in order of their FR numbers.

Class D5. 0-6-0 tender engines

The final development of the FR six-coupled goods engine came in 1913 when Pettigrew put into service four large-boilered mineral engines, with 4 ft 7½ in. wheels and standard 18 in. x 26 in. cylinders. Boiler pressure was raised to 170 lbs., and an extended smokebox, resting on a

D5 locomotive No. 32 with a trainload of coke from Bishop Auckland for Barrow picked up at Tebay passing the Dillicar Troughs south of the village. *John Alsop collection*

saddle, was fitted. As much as possible of the rest of the design was made standard with the earlier engines, though the frames were slightly thicker. There was also a tank version, Class L4. During the war years, the class was multiplied, fifteen further examples being built between 1914 and 1920, mainly by the North British Locomotive Co., though four built in 1918 came from Kitson & Co. All the 1913 engines were built with Phoenix smokebox superheaters (along with two 4-4-0 passenger engines). The apparatus was not a success and was removed in 1914. The appearance of the front end was considerably marred by the superheater, since an abnormally long smokebox was necessary, with the chimney placed right at the front, giving them a most ungainly look. When the apparatus was removed, normal extended smokeboxes were fitted.

The largest and most powerful of the mineral engines, this class did sterling work over the main line with the heavy iron-ore and coke trains, and later were often used for special excursion work, like all the other 0-6-0s being fitted with vacuum brakes and steam heating. In spite of their small wheels, they were capable of a good turn of speed when required. Some of the class had a very long life, surviving well into Nationalisation; the last to be withdrawn was 12510 (FR 33) in August 1957. On the other hand, four were scrapped as early as 1930, being then only twelve years old.

The later fifteen had boilers six inches longer than the first four, and in consequence had slightly larger heating surface, and weighed two tons more. There was a corresponding difference between the two pairs of 0-6-2 tanks which constituted Class L4. Tenders were the largest hitherto employed and were standard with the large 4-4-0 passenger engines of Class K4, built about the same time. These tenders held 3,300 gallons of water and 5 tons of coal, and in working order weighed slightly under 37½ tons.

The LMS renumbered the class 12494-12512, in order of building date. An interesting detail was that when 12509 was scrapped at Horwich Works in 1956, its tender, standing in the yard, bore distinctly discernable traces of all the liveries it had carried throughout its career. Unfortunately, these details would not show on a photograph and could only be seen when the light was caught in a certain direction across the side panel, but they were distinctly there: the British Railways lion and wheel (which one eminent authority, who shall be nameless, always referred to as "ferret and dart board"), underneath this the letters LMS; then the large figures 12509, and lastly, very faint under the layers of paint, the letters FR could just be distinguished.

Class E1. 2-4-0 tender engines

The increased weight of main line passenger trains was getting beyond the capacity of the little single-driver well-tanks, and a new class of 2-4-0 tender engines was ordered from Sharp, Stewart & Co. to replace them. The first pair, numbered 1 and 2, were delivered in 1870, and in the next three years fifteen further examples of the class were added to stock. After a gap of nine years, the final pair were delivered, one of which survived to become LMS property. They were a very simple design, and though small by modern standards they could achieve any amount of hard work. A number of exactly similar engines ran the passenger services of the Cambrian Railways, over a much more difficult terrain than the Furness, well into the 20th century. All the FR engines were later fitted with the automatic vacuum brake, and some of them were rebuilt with new boilers, having flush fireboxes, during the latter 1890s. In 1891, seven of them were given a new lease of life by being converted to 2-4-2 side-tanks for branch line service. The rear frames were lengthened and supported by a radial axle, a bunker, and side-tanks holding 1,000 gallons were added. Most of them survived the First World War, and two became LMS stock, though not for long, one being withdrawn in 1923 and the other in the following year. Neither actually carried the LMS numbers allotted. On rebuilding to tank engines, these seven were reclassified J1 and given a new diagram.

Much of their time as tank engines was spent in working the Lakeside branch and the through local services between Grange and Morecambe

E1 locomotive No. 12.   *John Alsop collection*

J1 locomotive No. 74. Formerly an E1 it was rebuilt in 1891 to become a 2-4-2 tank engine.
John Alsop collection

or Kendal. Two of them gravitated to Cleator Moor and worked the Joint Lines passenger services for several years. The tender engines began to be withdrawn in 1907, when two of the 1873 batch disappeared, but seven years elapsed before any more went to the scrap yard. The last two to be built, 44 and 45, were reboilered in 1898 and 1891 respectively, and survived the longest. Six, Nos. 1-4, 44, and 45 were put on the A list at various intervals, and 4A had the distinction of lasting no less than 13 years on the duplicate list. No. 44A was allotted 10002 in the LMS list, but never carried the number, being withdrawn in 1924.

The tenders attached to the 2-4-0s were of the same four-wheeled design as those of the D1 0-6-0 goods engines, holding 1,200 gallons of water and 2 tons of coal, weighing in working order 20 tons. Springs were above the running plate and the toolbox was placed at the rear, over the buffer beam. There were minor differences in the shape of the plate frames in different batches and also in the layout of the brake gear.

Class F1. 0-4-2 tender engines

Two engines of this class were taken over from the Whitehaven & Furness Junction Railway in 1866, on which line they were Nos. 3, *Mars*, and 13, *Sirius*. Built by R. & W. Hawthorn & Co. in 1857, they carried makers' Nos. 997/8, and were outside-framed engines with small four-wheeled tenders. They had peculiar dome casings, reminiscent of a large stew-pot with separate lid, even to the knob on the top; this feature was perpetuated by Hawthorn's from the middle 1840s for some 15 years on engines of their own design. Sometimes the knob on the top was omitted. Two very similar engines were built by the same firm in 1856, works Nos. 975/6; these were 1, *Excelsior*, and 2, *Hecla*, which passed to the LNWR on the split-up of the W&FJR stock. The Furness pair were renumbered 44 and 45, and continued to work in their old haunts until 1882, when they were both scrapped. They were small engines, built primarily for mixed traffic work, on passenger trains, and the lighter goods trains of which there were never very many. They had typical bell-mouthed chimneys, raised fireboxes, and Salter valves. The inside cylinders, 14 in. x 20 in., were under the smokebox and low slung, driving upwards onto the second axle. Coupled wheels were 5 ft diameter and the trailing wheels 3 ft 6 in. Axle guards were separate from the sandwich frames and were bolted on outside; the outside cranks were of the slab type. No cab was fitted, only a small front weatherboard. Owing to their limited capacity, it is surprising that the FR kept them in service for 16 years. The tenders were very small and had wooden frames, with axle-guards bolted on outside. The wheelbase was 8 ft 6 in., and they had a capacity of 1,500 gallons of water and 2h tons of coal, the weight in working order being 17 tons 7 cwt.

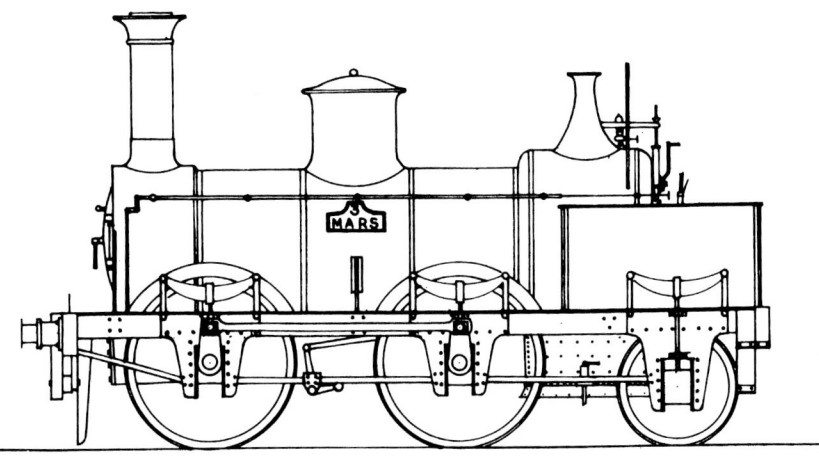

F1 locomotive as No. 3 *Mars* of the W&FJR.   *John Alsop collection*

Class G1. 0-6-0 side-tank engines

Lindal Bank, both sides of it, was causing some trouble to the Furness motive power department in the mid-1860s, with the heavy iron-ore trains which were destined for Lindal sidings. Something more powerful than the Bury 0-4-0s was necessary. Sharp, Stewart & Co. submitted a design for a heavy 0-6-0 side-tank engine with 4 ft 6 in. wheels and inside cylinders 18 in. x 24 in., which was accepted. The first two engines were delivered in 1867. They quickly proved they could tackle the work with success, but a falling off of ore traffic, due to one of the periodic recessions in the iron trade, did not make it necessary to order any further engines immediately. However, trade improving in the early 1870s, four more of the same class were ordered, two in 1872 and two in 1873. On the Furness system they were known to all and sundry as 'Neddies'. They were ungainly looking machines, with very long side-tanks which reached almost to the front of the smokebox, and had a flared coping. For attention to the motion, there was an oblong hole cut out of the bottom of each side-tank, centrally between the leading and driving axles. The dome was on the centre ring of the boiler and safety valves in a brass casing over the raised firebox. The cab was Sharpe's usual bent sheet of iron, strengthened with T-strapping. In later years, one or two engines had crude side sheets fitted to the cab, being merely pieces of wood planking bolted on with angle brackets. It is thought that this was an unofficial product of Moor Row shed, for those cabs must have been extremely uncomfortable out on the wind-swept

G1 'Neddie' locomotive No. 82. *John Alsop collection*

mountains of the Joint Lines around Lamplugh. No attempt was made to fit the engines with proper side sheets, so in all probability these home-made efforts had to be removed before the engines were sent in to Barrow for their periodic shopping. As the 'Neddies' spent a good deal of their time standing out in the open at Plumpton and Park South, waiting for trains to assist up Lindal Bank, it seems strange that no thought was given to the comfort of the enginemen in these horribly draughty cabs, but perhaps there was a handy shunters' hut.

The 'Neddies' never went far away from Lindal, except for two which put in a considerable spell in the Cleator Moor area, and consequently none of them achieved a very great mileage. In 1915, No. 52, then the only survivor of the first pair, was renumbered 84, but this was the only case of renumbering carried out on this class. The last four, of 1872-73, all survived into LMS days, but the class was extinct by 1925, and it is doubtful if any of them ever carried the LMS numbers allotted.

Class G2. 0-6-0 saddle-tank engines

Most of the Whitehaven, Cleator & Egremont engines are classified together, though they varied in details, makers, and age. In the main, they were all outside-framed engines, with 17 in. x 24 in. cylinders and 4 ft 6 in. wheels, with short saddle-tanks which covered the boiler only. No. 6, *Parkside*, was a distinct oddity in that as built she had a large all-over cab

G2 locomotive while still owned by the WC&ER No. 6 *Parkside*. *John Alsop collection*

with side windows, of distinctly American flavour, and also had the bottom row of tubes extended right through the front of the smokebox. This was said to "improve the draught", but whether it did so is a debatable point. Nos. 1, 2, 4, and 5 were domeless when built, but were fitted with domes at Moor Row when they came in for their first major overhaul. *Newton Manor* (No. 11) was an unlucky engine from the start; she had something not quite right with her trailing axle-boxes and was continually coming off the road, in consequence spending a considerable amount of time in Moor Row shops. However, in 1880 Mr Rose rebuilt her rear axle entirely, and thereafter she had no more trouble. This was after the engine had really distinguished herself with a boiler explosion at Seaton. The crew had left the engine, the driver to obtain the staff for the single-line section, and the fireman to couple up the train. There was a loud report and the boiler burst, sending clouds of steam and boiling water all over the place and blowing out most of the ballast. Luckily, the main blast seems to have been downwards. When the saddle-tank was lifted off in the works, it was found that the middle ring of the boiler had fractured just below the clack box. The engine was given a new steel boiler with 150 lbs. pressure (the original pressure was 125 lbs.) and was entirely rebuilt. The design was the basis of the rebuilding of all the engines of the class between 1880 and 1895.

Nos. 1, 2, and 4 were slightly smaller than the rest, having 16 in. cylinders. On rebuilding they were brought into line with the others, receiving new 17 in. cylinders and the 150 lbs. boiler. Though the rebuilding was in the main the same for all engines, there were some minor differences. The general pattern was for a flat-sided saddle-tank with semi-circular top (this was the original design), an all-over cab, in which the roof was continuous with side sheets and had a rain gutter sloping downwards on each side, from front to back, and a bunker with curved-up side panels and straight backplate, rather like the old standard GWR design. On most engines the saddle-tank covered the boiler only, but one or two engines had tanks which extended to the front of the smokebox. Nos. 5 and 10 varied in having completely semi-circular tanks, and the latter was also rebuilt without a cab, though it received one later. There were also variations in the shape of the bunkers. On the FR diagram they were all credited with the same tank capacity (1,000 gallons) but in all probability they actually varied between 900 and 1,100 gallons.

Little definite information is available about the actual dates of rebuilding, but No. 11 was the first, in 1880, and No. 4 was rebuilt in 1882 and again in 1895. When the engines came into Furness ownership in 1878, they were allotted Nos. 98-107, 110, and 111. Most of them were put on the A list when they were displaced by new 0-6-2 tanks, but several remained doing useful work in their old haunts through the First World War. All except No. 4, which came from Fletcher, Jennings & Co., of Whitehaven, were built by Robert Stephenson & Co. between 1855 and 1873. No. 109A was still in service at the Grouping, and was allotted LMS No. 11547, which it never carried, the engine being broken up in 1925, still in Furness livery.

Class G3. 0-6-0 saddle-tank engine

This was one of the four "odd" engines in the WC&ER stock. Though similar in most respects to the foregoing class, this engine, No. 112 in the FR list, differed in having inside frames. It was built by Andrew Barclay, Sons & Co., of Kilmarnock, in 1875, and carried their works No. 154. The heating surface was a little less than the Stephenson engines, but most other dimensions were the same. On the WC&ER, it carried No. 17, and was named *Wastwater*. In 1904 it was renumbered 108, and became 108A in 1907 when replaced in the capital list by a new 0-6-2 tank. It was rebuilt in 1896 and again in 1915, and on coming into the possession of the LMS was allotted 11548. In 1925 it was withdrawn and scrapped, the last survivor of the WC&ER stock.

G3 locomotive No. 108A, ex-WC&ER.   *John Alsop collection*

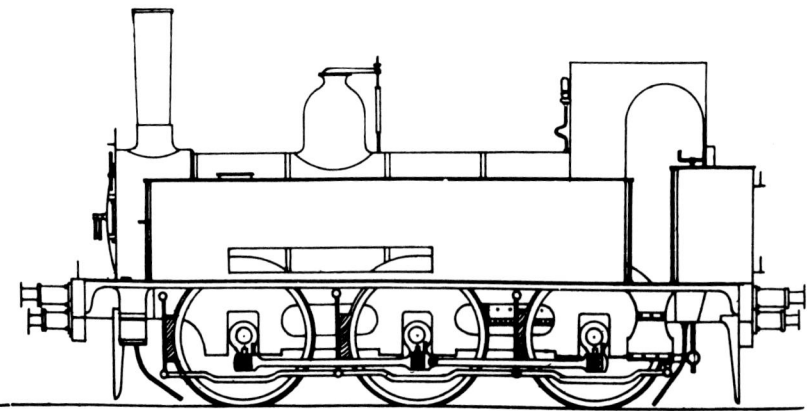

G4 locomotive, ex-WC&ER No. 3 *Victoria*. *Oakwood Press*

Class G4. 0-6-0 side-tank engine

No. 3, *Victoria*, was the other odd six-coupled tank engine, differing completely from the remainder. Built by R. & W. Hawthorn in 1857, works No. 989, she was a side-tank with typical features of the makers' designs. The FR renumbered her 113, and she was never on the A list, being sold in 1898. The side-tanks were long and had a narrow slot cut out of each side to allow access to the motion, a very similar arrangement to the 'Neddies'. As built, there was a stovepipe chimney and a dome of the stew-pot type, so typical of Hawthorn's. Only a front weatherboard was fitted, the footplate being entirely open. When rebuilt at Moor Row in the early 1880s, a normal dome, with Salter valves, was fitted, and a primitive all-over cab which had a semi-circular top to the cut-out. The stovepipe chimney was retained. The new boiler fitted at the same time was of similar dimensions to the original, but had a flush firebox. This engine had 14 in. x 22 in. inside cylinders and 4 ft 6 in. wheels; the side-tanks held 800 gallons.

Class G5.0-6-0 side-tank engines

To replace the 'Neddies', which were getting past their prime, and also for shunting and minor goods duties, Pettigrew evolved his own design of 0-6-0 tank engines. The first batch of six were built by the Vulcan Foundry (works Nos. 2523-28) in 1910, and carried running Nos. 19-24.

In 1918 they were renumbered 55-60 to bring them into the same sequence as the later batch. For a time they were used for shunting in Barrow Docks and also did a turn on the banking duties around Lindal, but latterly a couple were shedded at Moor Row for passenger duties on the Joint Lines. During the war years, when locomotive power was at a premium and some of the older stock was worn out but still doing useful work, four more of the class were ordered. These were Nos. 51-54, of which the first pair were built by Kitson & Co. (works Nos. 5121/2 of 1915) and the second pair delivered in the following year by the Vulcan Foundry (3174/5). These last four differed in minor respects from the original batch and were recorded on a separate FR diagram. The boiler was pitched two inches higher; the bunker backplate was a different shape, and the cab cut-out had a straight top edge. Water capacity was increased by 20 gallons and coal by 5 cwt.; and in consequence they were 13 cwt. heavier.

All ten engines were equipped for passenger working, though apart from the WC&ER line and occasional trips on the Coniston and Lakeside branches, they were not called upon much for such services. They were neat engines, and had Pettigrew's standard extended smokebox, supported on a saddle, and the typical Sharp Stewart design of cab, with curved top edge, which he also adopted as standard.

In 1923 the LMS renumbered them in date order 11553-62. The drastic reduction in trade caused by the great slump of the early 1930s took its

G5 locomotive No. 56. *John Alsop collection*

toll of this class, for between 1930 and 1932 six of them were withdrawn, though far from worn out. One each went in 1934, 1935, and 1936, leaving 11553 as the sole survivor, which was withdrawn in 1942 and cut up at Horwich in the following year.

Class H1. 2-4-0 side-tank engine

No. 12, *Marron*, was the third odd engine in the WC&ER list. It was purchased second-hand in 1870 from the North London Railway, for whom it was built as No. 3 in 1850 by Stothert & Slaughter, of Bristol. It was disliked by most of the WC&ER crews as an ungainly brute, and its 5 ft 6 in. coupled wheels hardly made it suitable for the mountainous gradients of the Joint Lines, even with the light trains of that period, and in 1872 it was sent to Fletcher, Jennings & Co. to be fitted with new 5 ft wheels and 15 in. cylinders in place of the original 14 in. x 20 in. At the same time, a cab of hideous design was fitted. For a long time after its acquisition by the WC&ER it retained the old North London Railway brass number 3 on the front of the chimney. The outside cylinders were placed in front of the leading axle, giving the engine a distinct and uncomfortable nosing action except at very low speeds. Her career after the Furness Railway took over was not very long; she was renumbered 108 and was scrapped in 1898, unwept, unhonoured, and unsung, and, it must be said, to the great relief of the crews who had to put up with her.

The fourth odd engine in the Cleator list was No. 16, and this did not come into Furness ownership, being handed over to the LNWR, who used it for some considerable time as a shunting engine at Crewe Works,

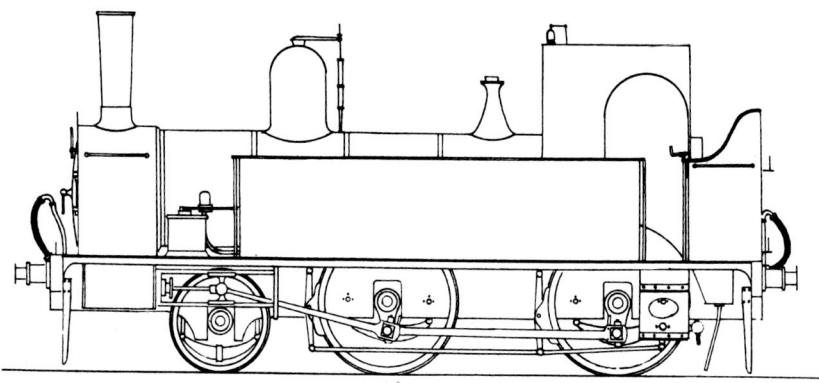

H1 locomotive ex-WC&ER No. 12 *Marron*.   *Oakwood Press*

and it was not given a number in LNWR stock. Eventually, in 1916, it turned up in the yards of the Whitehaven Iron & Steel Co., and three years later had gone to the scrapheap. *Ullswater*, as the engine was named, was an 0-4-0 saddle-tank of Andrew Barclay's standard type, works No. 153 of 1875, with 4 ft wheels and outside cylinders 14 in. x 22 in. Beyond this nothing is known of its dimensions.

As a general note, all WC&ER and W&FJR locomotives had a second pair of buffers at each end, placed inside and slightly below the normal set, for dealing with the small colliery chaldron wagons which were in general use in West Cumberland at that time.

Class K1. 4-4-0 tender engines

The arrangements with the Midland Railway over the Irish boat trains had necessitated some more powerful locomotives than the E1 2-4-0s which had been the maids of all work in this respect for some 20 years. In 1890, Sharp, Stewart & Co. were asked to tender for four passenger engines, and they submitted the design which had already been supplied to the Cambrian Railways, a 4-4-0 of moderate size. Accepted by the Furness board of directors, the four engines were delivered in 1891, numbered 120-123 and 3618-21 in the makers' books. Basically they were the 2-4-0s enlarged and fitted with a leading bogie; the

K1 locomotive No. 120.   *John Alsop collection*

coupled wheels were the same diameter, 5 ft 6 in., but the cylinders were enlarged to 17 in. x 24 in. The boiler was the same diameter but slightly longer, and the usual Sharp firebox was fitted. When delivered, the bogie wheels were fitted with small independent splashers, but these were subsequently removed, having caused some slight trouble on sharp curves, particularly in shed yards.

Though a considerable advance on the 2-4-0s, the 'Seagulls', as they were nicknamed, were soon themselves outclassed and were relegated to secondary duties. All were still in service in 1923, and were renumbered 10131-4, in original order, by the LMS. All four were repainted in LMS red livery, with large figures on the tender, and were withdrawn in 1927, 1928, 1927, and 1925 respectively.

The tenders fitted were the only ones of this particular type. They were virtually an enlarged version of the standard four-wheeled tenders but running on six wheels with an equally divided wheelbase of 11 ft 6 in. They had the same plate framing, with springs above the running plate, and a narrow body. Their capacity was 2,500 gallons of water and 4 tons of coal, weighing in working order 28¼ tons. They were the first tenders to be built with iron brake blocks.

Class K2. 4-4-0 tender engines

To supersede the 'Seagulls' on the heavier and more important trains, Sharp, Stewart & Co. supplied a class of 4-4-0 from an anonymous design by James Manson, of the Great North of Scotland Railway. Six were built in 1896, considerably larger than the previous locomotives of the type, with 6 ft coupled wheels, the largest hitherto used on the line- and 18 in. x 24 in. cylinders. The boilers had flush fireboxes, and again were the largest so far used. Two further engines were added in 1900, though in this case urgency of delivery was the main consideration, and the order for six engines of the next class (K3) was amended to four K3s and two K2s in order to get them into service as soon as possible. In 1913, two of the class, 37 and 34, were fitted experimentally with Phoenix smokebox superheaters, along with four 0-6-0s of Class D5 (then under construction). This apparatus, with its greatly extended smokebox and chimney perched right forward, completely spoilt the neat appearance of the engines. Moreover, it did not prove much advantage either to the running or steaming qualities and was removed from all six engines in 1914. Thereafter the Furness never had any superheated engines.

K2 locomotive No. 35. *John Alsop collection*

All eight of the class were in service in 1923, and were renumbered in the LMS series 10135-42. They lasted until the late 1920s, on secondary duties on their home ground. A considerable amount of renumbering took place in Furness days, making the numbering of this class somewhat complicated.

The tenders fitted to this class were the first of the modern design to be built, and were also fitted to the 0-6-0 goods engines of Class D3. Straight side panels with flared tops were used, and there were no coal rails, though some tenders had them fitted later. Springs were below the running plate, with splayed anchor links, thus for the first time allowing a full-width body to be used. They ran on six wheels with an equally divided wheelbase of 12 ft, and carried 2,500 gallons and 3½ tons of coal, weight being 28¼ tons. Twenty of these tenders were built, eight for the 4-4-0s and the rest for the D3 goods engines.

Class K3. 4-4-0 tender engines

Mr Aslett's determined efforts to improve and widen the scope of the passenger services bore fruit rather quicker than had been anticipated, and in 1900 the company again found themselves short of passenger motive power. Having already designed a new 0-6-0 for goods service,

Mr Pettigrew turned his attention to the other side and produced an enlarged 4-4-0 of his own. Six were ordered from Sharp, Stewart & Co., but such was the urgency for more power that the first two were changed to the K2 class, since this being a repeat order, for which all patterns and drawings were already available, delivery could be expedited. The four engines of Class K3 were delivered in 1901 and became the premier express engines on the line until, in their turn, they were displaced by Pettigrew's ultimate design of 1913. The K3s were based on the previous class and had a number of their characteristics in deference to a degree of standardisation. Their coupled wheels were increased to 6 ft 6 in. and cylinder stroke was increased to 26 in., the 18 in. diameter being unchanged. A much larger boiler was employed and the separate coupling rod splashers of the previous two classes were dispensed with. Handsome and efficient engines, two of them were stationed at either end of the main line, Carnforth and Whitehaven, from where they worked all the main line express traffic for over twelve years. Even when displaced by the very large K4 engines in 1913, they still took an odd turn or so on the fastest traffic and dealt with most of the excursion trains. Had the great slump of the 1929-35 period not upset traffic all over the country, it is quite likely they would have lasted much longer, but, becoming redundant by the reduction of passenger services, they were withdrawn in 1930-31.

On delivery, they came into the Furness list as Nos. 126-129, Sharp's works Nos. being 4716-19 of 1901. In 1923 they became LMS 10143-6, in

K3 locomotive No. 129. *Oakwood Press*

K3 locomotive No. 128 at Barrow Locomotive Sheds.   *John Alsop collection*

original order, the last three being withdrawn in 1930 while 10143 lasted a year longer.

The tenders fitted to this class were also used for the four 0-6-0s of Class D4. They were similar to the previous design, but were fitted with coal rails from the beginning, and the wheelbase was one foot longer, 13 ft The springs were below the running plate but, unlike the K2 tenders, had vertical anchor links. The capacity was increased to 3,000 gallons of water and 5 tons of coal, and weight in working order was 32 tons 14 cwt.

Class K4. 4-4-0 tender engines

Pettigrew's last passenger engines were contemporary with his large goods class, two of each being built as part of the same order in 1913. Though very large machines, they were not superheated. The boiler was standard with that of the goods engines (D5). The coupled wheels were reduced to 6 ft 0 in., but the cylinders remained at 18 in. x 26 in. With an increase of boiler pressure of 10 lbs. over the K3s, they were nominally only slightly more powerful, but their much larger boilers could cope with faster and heavier running with ease. Owing to their weight, these engines were not often used beyond Barrow, where two of them were shedded, the other pair being at Carnforth. Their chief duties were on

K4 locomotive No. 131 at Carnforth station. *John Alsop collection*

the through Barrow to Euston trains, which they worked as far as Lancaster. The 1913 and 1914 pairs differed slightly in dimensions, corresponding with the D5 0-6-0s built contemporaneously. The K4s were Pettigrew's magnum opus; handsome and imposing engines, even larger than the contemporary LNWR designs, thus adding something to the prestige of the small company. A comparison of a photograph of these locomotives with one of an Adams T6 of the London & South Western Railway betrays the fact that apart from the T6 being outside cylindered, there was a close resemblance between the two. In the same way, Pettigrew's D5 goods engines were very similar to Adams's "main line goods" class. Pettigrew, as has been stated already, was Adams's assistant at Eastleigh before he joined the Furness Railway.

These useful engines succumbed to the ravages of the great slump, aided by the LMS policy of getting rid of small classes (numerically) which were not standard, and all four were withdrawn in 1932. In fact, as far as the LMS was concerned, the years 1930 to 1932 were extremely bad for 4-4-0s. This period saw the demise of most of the numerous L&YR examples of the type, all the Furness, and large numbers of the older LNWR and Scottish engines of this wheel arrangement. Only those of the Midland flourished.

The four K4 4-4-0s were Furness Nos. 130-133. The first two carried North British Locomotive Co. works Nos. 20071/2 of 1913 and the other pair 20867/8 of 1914. Their LMS Nos. were 10185-88, coming between the 7 ft 3 in. L&YR 4-4-0s and the superheated version of the same class.

FURNESS RAILWAY LOCOMOTIVES

Class L1. 0-6-2 side-tank engines

Although the WC&ER saddle-tanks had done good work in West Cumberland, something larger was clearly necessary, for the traffic was getting too heavy for them to handle singly. Pettigrew therefore designed an 0-6-2 tank to take over the heaviest duties. It is probable that a good deal of the detail was left to the builders, for the engines bore a remarkable similarity to those built contemporaneously by Sharp, Stewart & Co. for the coal traffic on the Barry Railway in South Wales. Three engines were ordered and were delivered in 1898, numbered 112-14. They had 4 ft 7½ in. coupled wheels and 18 in. x 26 in. cylinders, and as much as possible was made standard with the tender engines of Class D3, which were ordered a few months later. The 0-6-2 tanks proved quite capable of the duties required of them, and remained in the Cleator Moor area for most of their life. The class was not repeated, though eventually there were four distinct classes of 0-6-2 tanks, this being the smallest. The makers' numbers were 4364-66, and on coming into LMS stock, Nos. 112-14 became 11622-24. The first (11622) was broken up in 1927 and the other two in the following year.

L1 locomotive No. 112. *John Alsop collection*

L2 Locomotive No. 104 on the Lakeside branch; most of the coach stock is six-wheeled.
John Alsop collection

L3 Locomotive No. 97. *John Alsop collection*

Class L2. 0-6-2 side-tank engines

A further batch of ten 0-6-2 tanks were built in 1904, five by Nasmyth, Wilson & Co. and five by the North British Locomotive Co. (as successors to Sharp, Stewart). These were very similar to the previous class but had 5 ft 1 in. wheels and were rated as mixed traffic engines. They were intended originally for banking duties, but their larger wheels made them suitable for passenger-service if so required, and they were fitted with vacuum brakes and steam heating apparatus. In most particulars they were standard with the 0-6-0 tender engines of Class D4. Being the largest class, numerically, of the 0-6-2s, they could be found all over the system, but their main spheres of operation were in the Barrow and Cleator districts. In 1923 they became 11625-34 in the LMS list. One, 11628, outlasted the bulk of the Furness stock, not being withdrawn until 1946. By 1936 there were only three 0-6-2 tanks left out of a total of 23; two engines of Class L3 were withdrawn in 1938 and 1941, but all the rest had gone by 1936. Nos. 11630 and 11633 received L&YR saturated Belpaire boilers in 1927.

Class L3. 0-6-2 side-tank engines

Six further examples of the type were added to stock in 1907, built by the North British Locomotive Co. In all main dimensions they were identical with Class L2, but had shorter side-tanks and longer bunkers; the tanks stopped short over the centre of the driving axle, and to provide the same water capacity the bunker tank was enlarged. In fact the only real difference between the two classes was the weight distribution. These six engines became LMS 11635-40 in 1923, and lasted rather longer than the L2s, probably on account of their different weight distribution, which made them rather more universally available. No. 11635, and possibly one other, was fitted with a L&YR Belpaire boiler in 1927.

NBL works Nos. were 17808-13 of 1907; the Furness numbers. were 96, 97, and 108-111 respectively, becoming LMS 11635-40, in that order. Two, 11638 and 11640, were withdrawn in 1933; 11637 in 1935; 11639 was the first to go, in 1931, but 11635 and 11636 remained in use until 1938 and 1941 respectively.

Class L4. 0-6-2 side-tank engines

The final class of 0-6-2 tanks were four in number, and were delivered in two pairs, which had some differences in dimensions. All were built by Kitson & Co., the first two in 1912 and the others in 1914. The latter pair had boilers standard with the D5 0-6-0s; the 1912 engines had boilers which were not standard with any other class, though five further boilers were built subsequently with the same dimensions, for fitting to the Sharp goods engines of Class D1. These boilers were 1 ft 3 in. shorter, the difference being made up by extending the smokebox backwards, which gave the chimney the appearance of being in the centre of the smokebox, while in the longer boilers the chimney appeared to be further back. The shorter boiler, together with only 208 tubes instead of 230, gave a smaller heating surface. The four engines of Class L4 were known as the "Improved Cleator Tanks" and, as implied, they spent most of their lifetime in that area. Like all the other 0-6-2s, except Class L1, they were fitted for passenger service. They were the most powerful of the L classes, but were also the shortest in overall length and had the least water capacity. In 1923 they became LMS 11641-44, and though by far the youngest in point of age, they were the first to be scrapped, 11642 in 1929, 11644 in 1932, and the remaining pair in 1934. Their coupled wheels were 4 ft 7½ in. diameter, cylinders 18 in. x 26 in., and the boilers were pressed to 170 lbs. In the FR list they were 92-95, of which 92/3 were Kitson's 5042/3 of 1914 and 94/5 were Kitson's 4855/6 of 1912. The LMS numbers were allotted in date order.

L4 locomotive No. 94. *John Alsop collection*

FURNESS RAILWAY LOCOMOTIVES 139

Class M1. 4-4-2 side-tank engines

For working the branch lines to Coniston, Lakeside, Kendal, and Morecambe, a new type of passenger engine was designed by Mr Pettigrew. Six were built, the first two, 38/9, by Kitson in 1915, two more by the Vulcan Foundry in 1916 (40/1), and the final pair (42/3) by Kitson, also in 1916. Cylinders and boilers were standard with the G5 0-6-0 tanks, for Nos. 38/9, but the other four had boilers with 220 tubes instead of 208, the heating surface being 1,039 sq. ft They were handy little engines and could perform the tasks required of them without fuss, both on the long 1 in 49 gradient of the Coniston branch and the almost level LNWR line into Lancaster, on which they produced a good turn of speed. When first put into service they did a good deal of work on the heavy and intensive workmen's services around Barrow, and it was not until 1919 that they were put to work on the branches for which they were intended. The LMS renumbered them 11080-85, and they were withdrawn from service between 1930 and 1932 after a short career, martyrs to the LMS policy of standardisation and removal of small classes.

M1 locomotive No. 38. *John Alsop collection*

Class N1. 4-6-4 side-tank engines

Mr Pettigrew retired in 1919 and his successor, Mr D. L. Rutherford, brought out the last design of express locomotive to be built for the Furness Railway. Having regard to the comparatively short time which elapsed between Pettigrew's retirement and the building of these locomotives, it has given rise to a considerable doubt as to who actually designed them. It is quite possible that the original conception was Pettigrew's but that it fell to Rutherford to actually get the project under way. To complicate matters still further, one authority states that the main details were worked out by the chief draughtsman at Barrow, one Mr Sharples. However, to whom goes the credit for these fine engines is not of very great importance; the fact that they were built, and were highly successful, is the main thing. Certainly the FR had never seen anything like them before; they were built to the extreme limits of the loading gauge; in fact, on the odd occasions when one had to stand overnight at Whitehaven, it had to be stabled in the open, since the shed doors could not cope with an engine whose chimney top was 13 ft 6 in. above the rails. The only inside-cylindered 4-6-4 tanks ever to run in Great Britain, they were also the only unsuperheated ones. The "Jumbos" as the staff promptly named them were very good engines, and after the initial teething troubles had been ironed out were

N1 locomotive No. 115 on a mixed train with two Great Central Railway vans closest to the engine. *John Alsop collection*

extremely popular with the crews. During her initial trials, No. 115 easily negotiated curves of five chains radius. Coal and water capacity was sufficient for the through run between Carnforth and Whitehaven. They were, however, prohibited from working north of Whitehaven, on either the Joint Lines or the C&WR, on account of their weight, not that there was any necessity for this prohibition really as they were far beyond anything in the motive power field required for these lines. Their chief duties were the up and down mail trains between Carnforth and Whitehaven and the up morning express from Whitehaven. They also took over the through Midland trains from Barrow, handing over to the MR at Carnforth East, as was the usual practice. They also appeared occasionally at Lancaster and Kendal, though not regularly.

Kitson & Co. were the builders of Nos. 115-119, the first four being delivered late in 1920, and No. 119 in January 1921. The design was clean and comparatively simple and produced one of the most handsome express tank engines of the period. They were the first and only Furness engines to have Belpaire fireboxes. There is one curious difference between the official FR diagram and the engines as built; the diagram shows the running plate level from the buffer beam to the rear of the smokebox before being raised over the coupled wheels, while in the engines the rise was at the front of the smokebox. Owing to the size and pitch of the boiler, the tanks had to be made rather narrow, holding only 1,475 gallons, and the balance of 1,325 gallons was held in a tank under the bunker. The absence of outside cylinders and valve gear enhanced rather than detracted from the appearance of the engines. In the LMS list they were numbered 11100-4. No. 11102, was withdrawn in 1934; 11100, 11101 and 11104 in 1935, but 11103 lasted until 1940. Kitson's works numbers for 115-119 were 5292-96.

Coupled wheels were 5 ft 8 in. diameter and bogie wheels 3 ft 2 in. The wheelbase was symmetrical, totalling 40 ft 9 in., of which the rigid wheelbase was 13 ft 6 in. The inside cylinders were 19 in. x 26 in. and the boiler pressure 170 lbs.

Unclassified: 0-4-0 rail motors

In 1905 Pettigrew designed and built at Barrow Works (the only engines to be actually built there) two four-coupled rail motor cars for use on the Coniston and Lakeside branches. The engine was enclosed within the car body, which was of metal construction for the engine compartment but of normal wood framing and panelling for the passenger portion,

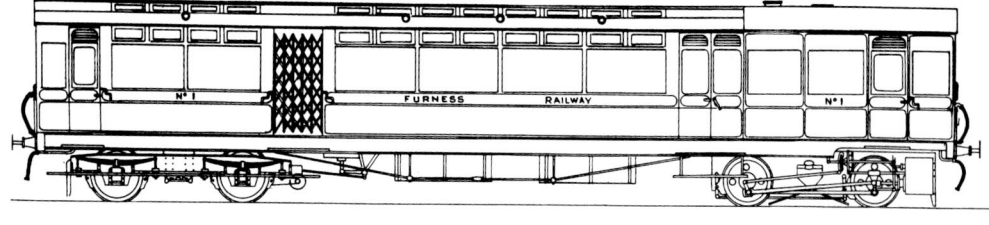

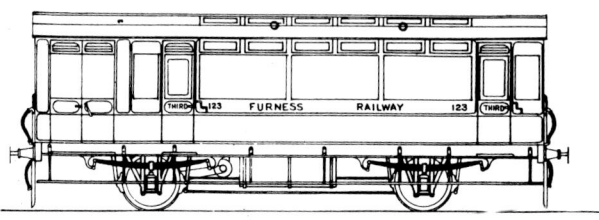

Rail motor No. 1 and trailer No. 123. *Oakwood Press*

which seated 12 first class passengers in the rear and 36 third class in the main saloon. Entrance to the car was by an open vestibule between the first and third class saloons; while running, a Bostwick steel gate closed the entrance on either side. At the extreme rear was a guard's compartment, which was fitted with control gear for reverse running.

The engine consisted of a small locomotive boiler, placed with firebox leading, supplying steam to a pair of 11 in. x 14 in. outside cylinders operated by Walschaerts valve gear. The cylinders were at the extreme front and drove onto the rear coupled axle. The engine compartment had an elliptical steel roof, but the passenger portion had a clerestory again an almost unique feature on the line. Two four-wheeled trailer cars, seating 28 third class passengers, were also constructed to run with the motors, these also fitted with control gear at one end. Both were later partitioned to give a small first class compartment 9 ft long.

On trial on the Coniston branch, the unit successfully negotiated the 1 in 49 gradient, fully loaded, and with trailer. It also successfully started again after being stopped on the same gradient. No. 2 was written off at an early date in a hushed-up encounter with a buffer-stop; the other was withdrawn after some nine or ten years' service on account of excessive vibration, a complaint which seemed prevalent amongst vehicles of this type. The motors were numbered 1 and 2 in a separate series, but the two trailers were numbered 123 and 193 in the ordinary passenger stock.

Rail motor No. 1 at Coniston station.   *John Alsop collection*

Rail motor No. 1 at Lakeside station.   *John Alsop collection*

FURNESS RAILWAY.
BARROW DOCKS

OWNED AND WORKED BY THE FURNESS RAILWAY COMPANY.

H.M.S. Princess Royal and Japanese Battleship Kongo, Buccleuch Dock.

THE THIRD LARGEST DOCKS IN THE UNITED KINGDOM.

Consist of the "DEVONSHIRE," "BUCCLEUCH," "RAMSDEN," & " CAVENDISH " Docks, comprising

299 ACRES OF WATER SPACE,

With an entrance 100 feet wide, and a depth of 31 feet 6 inches at H.W.O.S.

A DEEP WATER BERTH HAS BEEN CONSTRUCTED IN WALNEY CHANNEL, NEAR RAMSDEN DOCK STATION, 850 FEET LONG AND 100 FEET WIDE, WITH A DEPTH OF WATER AT L.W.O.S. OF 33 FEET.

The Entrance Basin at Ramsden Dock is 900 feet long. Petroleum Storage for 34,520 tons. | The Docks and Quays are lighted with electricity. Cranage Power up to 150 tons.

The Railway extends to all the Quays, and Wagons pass direct into and alongside all the WORKS and WAREHOUSES. Applications for Rates of Carriage and Freight, Dock and Labour Charges, etc., to be made to Mr. T. JACKSON, GOODS MANAGER, BARROW-IN-FURNESS ; and for Fares and Rates for Passenger Trains, Train Arrangements, etc., to Mr. A. A. HAYNES, SUPERINTENDENT OF THE LINE, BARROW-IN-FURNESS

Industrial Facilities at BARROW-IN-FURNESS

UNRIVALLED SITES FOR NEW WORKS, FACTORIES, WAREHOUSES, TIMBER AND OTHER WHARVES. **OVER 140 ACRES OF AVAILABLE LAND** ADJOINING THE FURNESS RAILWAY COMPANY'S EXTENSIVE DEEP WATER DOCKS, WITH RAILWAY ACCOMMODATION.

MANUFACTURERS, TRADERS, AND OTHERS are invited to apply to the GENERAL MANAGER, FURNESS RAILWAY, BARROW-IN-FURNESS. | The Company's NEW PLAN showing the available Sites on Barrow Docks, and in the town of Barrow-in-Furness, the property of the Furness Railway Company, will be forwarded on application, with full particulars.

Above: A Furness Railway advert promoting the Barrow Docks.
Below: A ship in a floating dock within the Devonshire Dock.   Oakwood Press

Chapter Nine

Whitehaven & Furness Junction Railway Locomotives

Though several W&FJR locomotives did not become Furness property, being handed over to the LNWR in the general share-out, a few notes on them may not be out of place. There is some doubt about certain engines, since most of them did not last long on the LNWR, in whose books they were given capital list numbers 1551-1560. Within a few months these numbers were altered to 1578-1587, and thereafter some of them had a variety of numbers in the duplicate list. As the LNWR list, particularly in the 1100 and 1200 series of duplicate engines, is often at variance with itself, the various engines concerned are difficult to trace with accuracy.

The first two engines on the W&FJR were Nos. 1, *Lowther*, and 2, *Whitehaven*, built by Tulk & Ley (forerunners of Fletcher, Jennings & Co.) in 1847, being Nos. 6 and 7 in the builders' list. They were outside-framed 0-4-2s with 4 ft 6 in. coupled wheels and 14 in. x 18 in. inside cylinders. No other dimensions are known. *Whitehaven* became 2A in 1856 (the only known use of a duplicate list on the W&FJR) and was scrapped in 1859. *Lowther* was broken up in 1856. No. 3, *Lonsdale*, was a 2-4-0 of R. & W. Hawthorn's build (No. 600) also in 1847, and was provided for passenger duties. This was also an outside-framed engine, with 5 ft 6 in. coupled wheels and 15 in. x 21 in. inside cylinders. It was withdrawn in 1857.

No further engines were added to stock until 1850, when two small 2-2-2 well-tanks, 4, *Oberon*, and 5, *Titania*, were bought from E. B. Wilson & Co. These became FR 47 and 48, and have been dealt with already under Class B5. In the same year, two further 2-4-0s were obtained, very similar to No. 3. There is some doubt as to the builders of No. 6, *Phoenix*. It is given in A. C. W. Lowe's list as Hawthorn, but without a works number, and cannot be traced in Hawthorn's list. There is a possibility that it was obtained second-hand. An old F. Moore photograph depicts it as a typical Hawthorn product of the late 1840s, and also shows it with outside coupling rods and cranks removed, so that at some period in its W&FJR career it ran as a 2-2-2. The coupled wheels were 5 ft 6 in. and inside cylinders 15 in. x 21 in. It passed to the LNWR as No. 1559 (later 1586) and with withdrawn in 1867. No. 7, *Petrel*, was very similar to *Phœnix*, except that the cylinders were 15 in. x 22 in., and was built by Stephenson's (No. 701) in 1850. It became LNWR 1556 and after various

145

Whitehaven & Furness Junction Railway locomotive No. 3 *Mars*.

John Alsop collection

Whitehaven & Furness Junction Railway locomotive No. 6 *Phœnix*.

John Alsop collection

renumberings was withdrawn in 1883. Two six-coupled goods engines followed, purchased second-hand from the Lancashire & Yorkshire Railway in 1854. Built by Hawthorn in 1846, works Nos. 466 and 465, for the Blackburn, Darwen & Bolton Railway, they became 217 and 218 in the L&YR list in 1849, and finally W&FJR 8, *Tubal Cain*, and 9, *King Lear*. Both were outside-framed engines with 4 ft 8 in. wheels and 16 in. x 24 in. inside cylinders; the boiler was 4 ft diameter and 10 ft long, with a total heating surface of 605 sq. ft The wheelbase was 14 ft 6 in., equally divided, and weight 23½ tons in working order. They passed to the LNWR and were renumbered 1558 and 1551, both being scrapped in 1872.

Two further engines were obtained from the L&YR in November 1854, 10, *Maryport*, and 11, *Kelpie*. They were inside-framed 0-4-2s, with outside cylinders 16 in. x 18 in., coupled wheels 4 ft 9 in., and were altogether a most peculiar design. The Vulcan Foundry built them in 1848 as part of an order for the Scottish Central Railway, but the order was reduced, and three of them (works Nos. 318-320) were sold to the Liverpool, Crosby & Southport Railway. The vicissitudes of this company are not for discussion here, but suffice it to say that the engines were twice in and out of L&YR stock within six years, and finally came to work in West Cumberland. No. 10 of the W&FJR was Vulcan No. 319. This engine did not last long enough to see the split-up of the stock, being withdrawn in 1860, but No. 11 (Vulcan 318) passed over to the LNWR as No. 1560 and was scrapped shortly afterwards. In L&YR stock, the two engines had been Nos. 224 and 223.

Tulk & Ley built the next engine in 1855, No. 12, *Big Ben*, which was an 0-6-0 goods engine with 4 ft 9 in. wheels and 16½ in. x 24 in. cylinders, beyond which nothing is known of its dimensions. It was No. 18 in the makers' list and became LNWR 1557. It was scrapped before 1870.

In 1856-57 four engines were added to stock, three of which were replacements of older engines withdrawn. These were Nos. 1, *Excelsior*, 2, *Hecla*, 3, *Mars*, and 13, *Sirius*, all built by R. & W. Hawthorn, works Nos. 975/6 of 1856, and 997/8 of 1857, in order. Of all the W&FJR locomotive stock, these were the only ones which could by any stretch of imagination be called a class, and their details will be found in the Furness Railway chapter under Class F1. Nos. 1 and 2, went to the LNWR as 1553/4 and were scrapped in 1874 and 1877 respectively. The other pair became FR 44 and 45 and were withdrawn in 1882. All were outside-framed 0-4-2s with 5 ft coupled wheels and cylinders 14 in. x 20 in.

Whitehaven & Furness Junction Railway locomotive No. 14 *Vulcan*.   *Oakwood Press*

The two 0-4-0 saddle-tanks which bore the numbers and names 15, *Bob Ridley*, and 16, *Banshee*, have been mentioned under FR Class C2; they became FR 49 and 50 and were both sold in 1882. Next in the list came an 0-6-0 tender engine, 17, *Garth*, which is also a bit of a mystery; it was ascribed by A. C. W. Lowe to Hawthorn's, with a date of either 1863 or 1864. However, it cannot be traced in that firm's list, and was possibly another second-hand engine. It became LNWR 1552, and after several renumberings was scrapped in 1886. Wheels were 4 ft 6 in. and inside cylinders 16 in. x 24 in. It may have been similar to the next three engines, which were of the 0-6-0 type, and had the same wheel and cylinder dimensions. These three, 14, *Vulcan*, 18, *Cedric*, and 19, *Lonsdale*, were designed by Mr Meikle, the locomotive superintendent of the company. Nos. 18 and 19, which became FR 43 and 42, have been dealt with under Class D2. No. 14 went to the LNWR as No. 1555 and was withdrawn at the early date of July 1867. It was built in 1860 by Hawthorn, works No. 1104.

Only one engine remains, and it replaced the original No. 10, which was scrapped in 1860. It was a 2-2-2 well-tank named *Queen Mab*. Its details and subsequent history will be found under Class B4 in the Furness Railway section.

These notes on the W&FJR locomotive stock complete the survey of the motive power of the Furness Railway. Though complicated in parts, owing to numerous renumberings, and with the paucity of reliable information on the W&FJR Stock, it is hoped that a reasonably accurate picture of the stock has emerged.

Chapter Ten

Rolling Stock

Passenger

That in the earliest days the company expected little passenger traffic is borne out by their order for only four coaches to meet the demands of the opening of the line. These were ordered through Edward Bury (who also supplied the first locomotives), but by whom they were actually built has not been ascertainable. Bury's works at Liverpool dealt solely with locomotives and did not undertake the construction of rolling stock, so they must have been contracted out, probably to some firm of coach-builders in the Liverpool-Manchester area. The four carriages were shipped by sea from Liverpool to Piel, along with sundry items of goods stock, in 1844.

These four coaches were typical of the period, with four spoked iron wheels running in plain bearings. They had three compartments, retaining the curved external mouldings of the "stage coach" pattern, and were very similar to a number of coaches built for the Liverpool & Manchester Railway about the same time. About 18 feet in length and 7 ft 3 in. wide, they seated nominally eight in each compartment, and were far from comfortable. One writer has described these coaches as a whole as "a model of elegant discomfort". A guard's seat was fitted at one end of the roof, which also had the usual rails for the transport of passengers' luggage. Lighting was by the inevitable noisome oil pots, one to each compartment, which though better than the usual system of making one lamp serve two compartments was still distinctly sepulchral. It is not certain how these coaches were arranged, but it is thought that two were entirely first class and the other two composites,

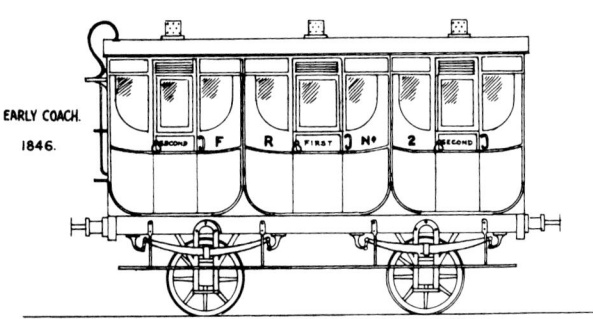

Furness coach No. 2, one of the early 'stage coach' patterns.   *Oakwood Press*

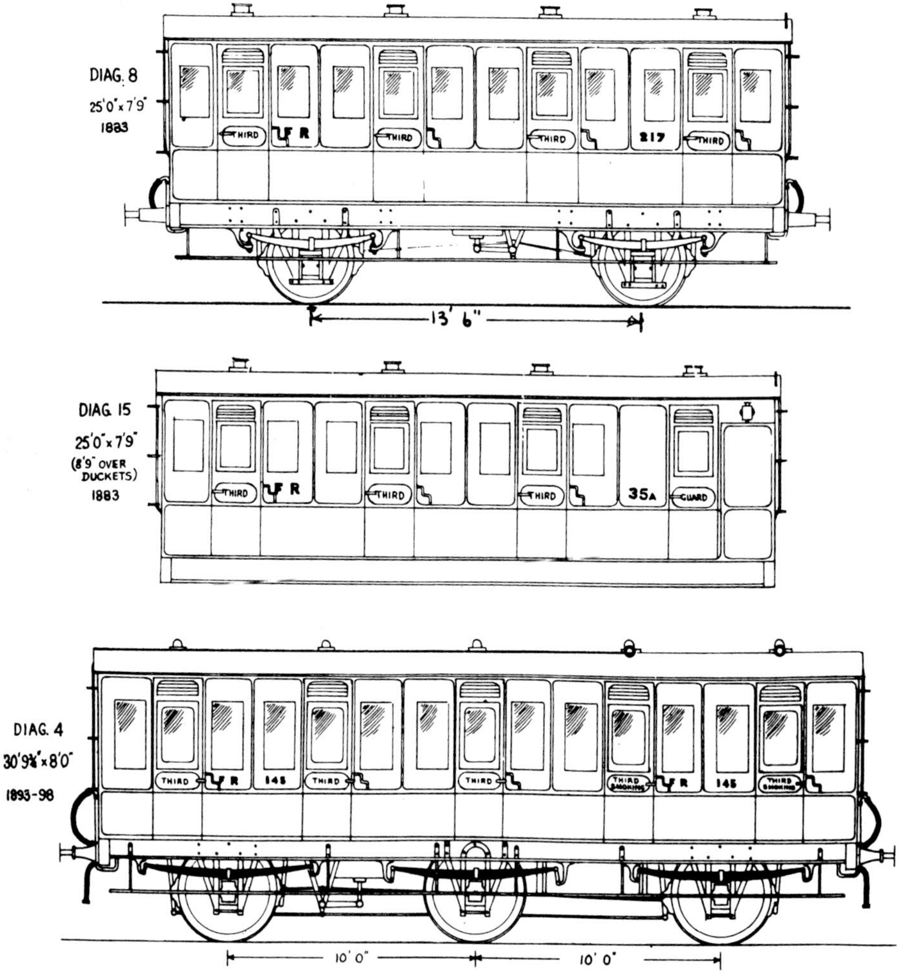

Furness Railway coaches of the 1880s.   Oakwood Press

Furness Railway train comprised of old four-wheeled stock with the Duke of Devonshire's saloon behind the engine.   John Alsop collection

with one first class compartment in the centre. Third class passengers were not catered for and were not even contemplated until 1850 when, in response to the Board of Trade's edict concerning "Parliamentary" trains, a few open wagons were fitted with seats of a kind. In later years, however, third class passengers became the main consideration, and entire first class coaches were abandoned after 1880, such traffic as the elite required being catered for by a variety of composite coaches. One or two specially posed photographs are extant, showing the entire stock of the early coaches, plus the Duke of Devonshire's special saloon, made up into a train headed by one of the Bury locomotives.

Addition was slow; by 1880 the total passenger stock still numbered less than 100, almost one locomotive for every carriage. The Furness Railway was still pre-eminently occupied with mineral traffic, and passengers were only a very secondary consideration. Until this time, the stock was almost exclusively four-wheeled, and third class five-compartment carriages, plus a few two-compartment brake thirds, were built down to 1885. A small number of decrepit four-wheelers, in assorted sizes, was inherited with the assets of the Whitehaven & Furness Junction Railway in 1866, but exactly how many is not known; they had all been replaced by the middle 1880s. After 1885, six-wheeled stock became the order of the day, until 1897, when the first bogie coaches were put into service, but the last six-wheeler was built in 1901.

The first six-wheelers were introduced in 1875, and one of these actually survived until after the Grouping, though rebuilt as a family saloon. However, the six-wheeler was a rarity until 1882, when a batch of six composites (six compartments) on 39 feet under-frames, were obtained. These vehicles were remarkable in that they had Cleminson patent radial axles; in this design the three axles were mounted as separate trucks, the centre one having a certain amount of lateral movement, in guides fixed to the under-frames. Radius rods fixed to this centre truck connected it with the two outer trucks, which were pivoted at their centres, thus the amount of swivelling was controlled by the movement of the centre axle. It will be seen from this arrangement that the total wheelbase had to be carefully calculated in relation to the sharpest curve which the vehicle would have to negotiate in order to obviate grinding of the flanges as far as possible. The first Cleminson coaches were built for the Woosung Tramway, in China, in 1875, and though quite a number of similar vehicles were constructed, they never became popular. More were in use on narrow gauge railways than on standard gauge; the Isle of Man and Southwold Railways, in this country, had a few, but the six composites of the Furness Railway are

Duke of Devonshire's saloon. *John Alsop collection*

Sir James Ramsden's saloon used as an inspection car. *Oakwood Press*

believed to be the most numerous batch of any built for the standard gauge in Great Britain. The riding was distinctly hard, and for this reason the design was not perpetuated. Two of the Furness composites were converted in 1893 to family saloons and were still in service in 1923. Of the other four, three were still extant at the Grouping, though out of use, and were scrapped very shortly afterwards.

In 1923 there were 144 six-wheelers still in stock, of which 107 were five-compartment thirds, eleven brake thirds, and the remainder composites of various types, all dating from the 1880s to 1901. The standard third class under-frame was 30 ft 9¾ in. over headstocks, but the composites varied in length between 33 ft 9¾ in. and 35 ft A small number of six-compartment thirds were built in 1884, 34 ft 3¾ in. long, but these had practically ceased to exist by 1923, only one being still extant at that date. The composites were a very mixed bunch; the Cleminsons have already been mentioned, these had two third class compartments at each end, then two second class compartments (later demoted to thirds) flanking two firsts in the centre. In 1889 four composite brakes were obtained, having two firsts and two thirds, plus a small van, on 34 ft 4 in. frames. During 1890-99, fourteen coaches with a third and first compartment at each end and a luggage compartment in the centre were obtained, 33 ft 10 in. over headstocks. Finally, in 1900-1, two semi-corridor composites were built at Barrow, with two lavatories in the centre and a luggage compartment at one end; these were 35 ft long.

Before leaving the older stock, two special carriages should be mentioned. The first of these was a four-wheeled vehicle built about 1853 for the private use of the Duke of Devonshire. It was a peculiar coach, with the panelling still adhering to the old stage coach tradition, and entrance by a narrow platform at one end only. It had a low elliptical roof and a type of clerestory with slats in the sides for ventilation. Unusual for the period were the three large side windows, which had bevelled edges, and in the opposite end to the platform were three round-headed windows. Inside, the coach was sumptuously furnished in brown leather, but the lighting gave a great deal to be desired, since there was but one solitary oil pot in the centre of the roof. The under-frame was definitely unusual, with trunnion bearings for the axles, reminiscent of Bury's practice for his locomotive tenders, with short springs inside the hangers and a strong tie bar between the axles. Some kind of brake gear was fitted, but it is not at all clear how this was worked. This coach was used extensively by the Duke for many years, finally being laid aside in the early 1900s, after which it languished for a time in Barrow works, whence it ultimately disappeared.

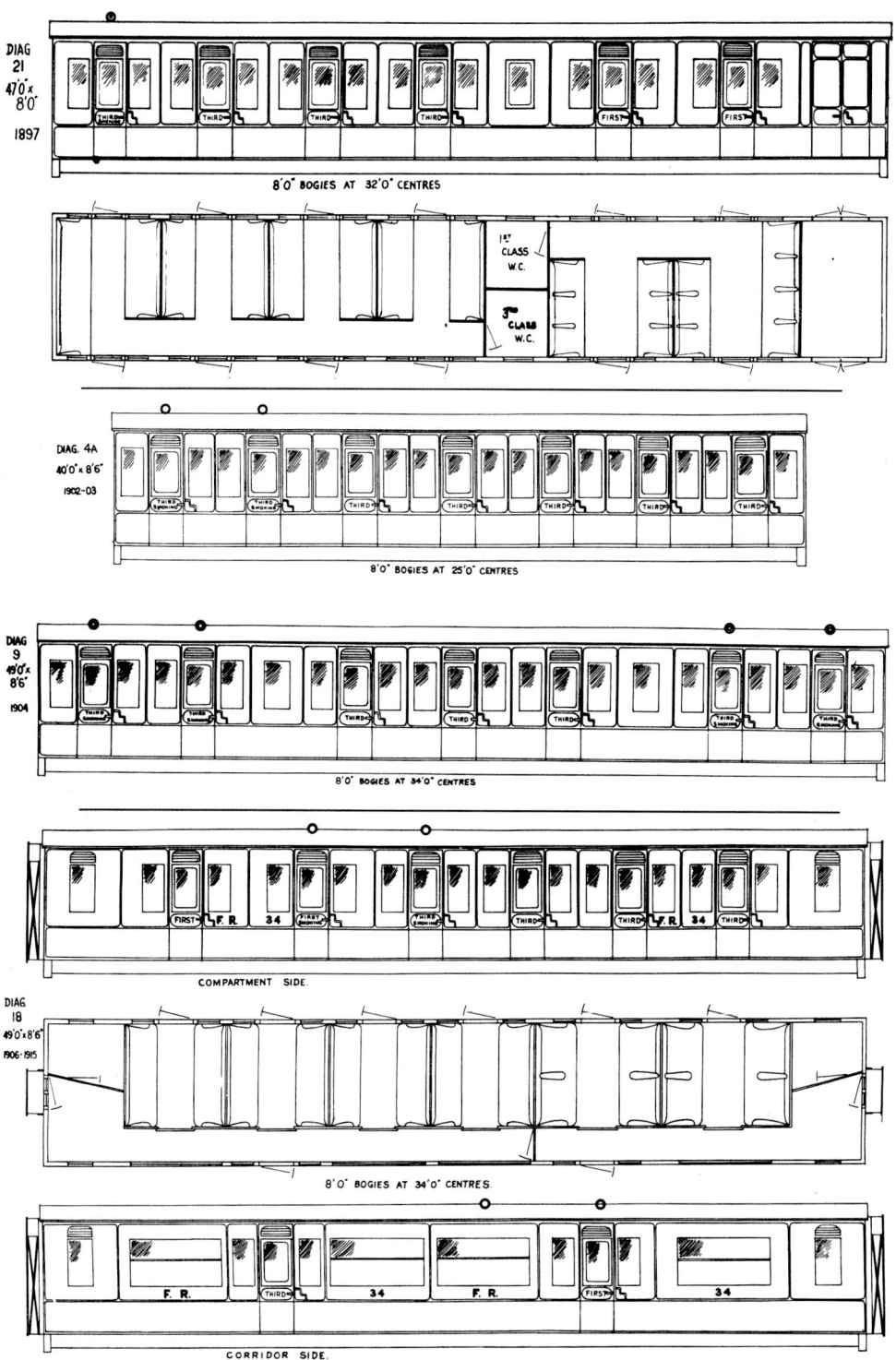

Furness Railway coach designs. Oakwood Press

The second vehicle was a special saloon built for Sir James Ramsden, and later used as a general inspection saloon. This was built by Wright Bros. of Birmingham (the forerunners of the Metropolitan Railway Carriage, Wagon & Finance Co.) in 1865. It was divided into two parts, of which the larger saloon was 9 ft 6in., later rebuilt a foot wider than the smaller section, which was 7 ft 6 in. in length. The extra width of the larger portion was designed to allow of observation both ways. At the rear end was a short open verandah with a well floor. The vehicle ran on four wheels, with a base of 9 ft (later 11 ft), and was upholstered in brown leather, but the motion, like that of the Duke's saloon, was horrible. The under-frame was of orthodox carriage pattern with long springs. In 1884 vacuum brake gear was fitted. This saloon was preserved; original livery was white upper panels, dark below waist, with two FRC monograms flanking an FR crest on the door. It was later (probably about 1884) painted Indian red all over.

The turn of the twentieth century saw more bogie vehicles coming into use; the first had appeared in 1897, in the shape of six semi-corridor composite lavatory brakes and six similar coaches with a luggage compartment in place of the guard's van, all on 47 ft under-frames. For local traffic, the six-wheeler still held sway, but some 48 ft eight-compartment thirds came into use in 1902, plus two short 40 ft seven-compartment thirds, one in the same year and the second in 1903. These two vehicles, with an inter-partition length of only 5 ft 6¾ in., were rather cramped, and no more were built. In 1904 appeared the first of a small number of lavatory thirds on 49 ft under-frames, and five 43 ft 6 in. thirds with only one lavatory were built between 1903 and 1907. The only other non-corridor thirds were twelve nine-compartment vehicles on 57 ft frames, built between 1919 and 1921. A number of four-compartment brake thirds on 49 ft frames appeared between 1906 and 1909, with an additional vehicle built to the same diagram in 1915. Eight six-compartment brake thirds, 48 feet long, had been added to stock in 1901-2.

The twelve bogie composites of 1897 comprised six with three thirds, two firsts, and guard's van; these had side-by-side lavatories between the first and third accommodation, thus serving only one compartment each. The other six were similarly arranged, except that they had an extra third class compartment. Eight brake composites without lavatories (a third, two first, two third, guard) followed in 1901-2, on 48 ft frames. The final batch of six composites came out in 1903. These were on 49 ft under-frames and had lavatory accommodation for first class only, the arrangement being two third, two first, two lavatory, one first, two third. Corridor stock with end gangway connections first appeared in 1904 for use as through coaches to London, Manchester, Leeds, etc. Only fourteen such vehicles were built, four between 1904 and 1914, four in 1914-15, and six between 1921 and 1924. All had side corridors and end vestibules, with a lavatory at each end. Four were seven-compartment thirds on 49 ft frames; four more were composites on the same under-frames, with three firsts and four thirds, while the post-war batch, which were the longest vehicles on the line (60 ft 7½ in.) reversed the arrangement, having four firsts and three thirds. These, along with the 57 ft nine-compartment thirds at 9 ft, were the widest stock the Furness ever had.

Seven miscellaneous saloons completed the passenger stock, and were numbered in a separate series, 1-7. Two of these, 4 and 5, were the

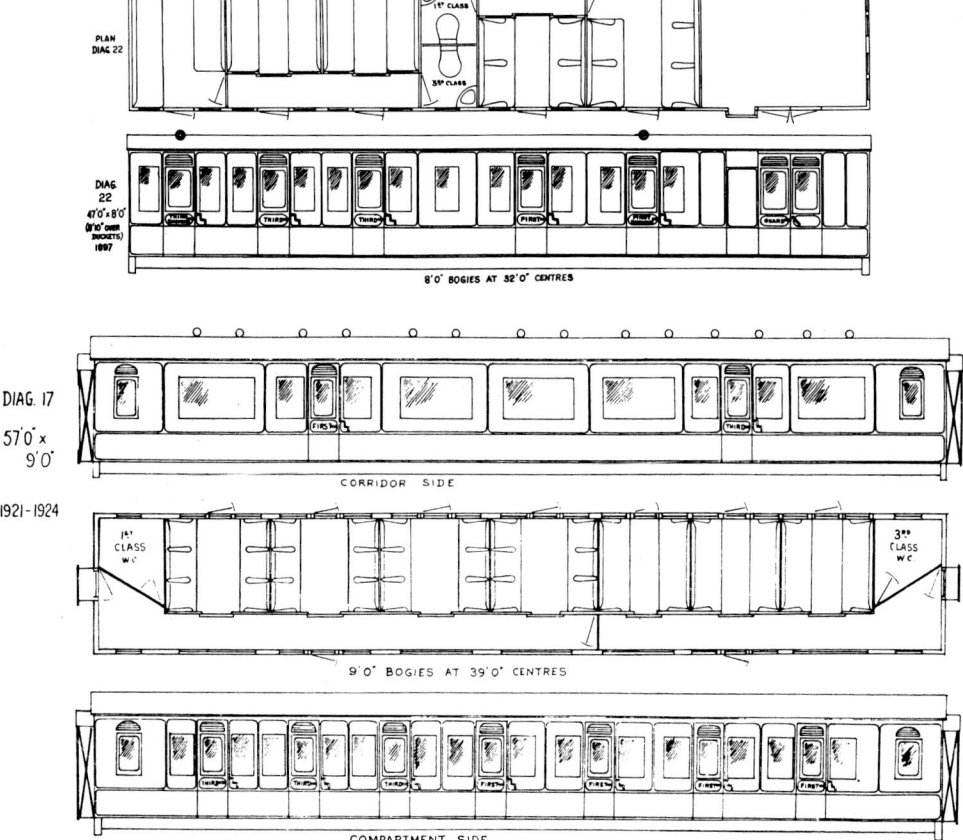

Furness Railway coach designs. *Oakwood Press*

Furness Railway saloon No.1. *John Alsop collection*

conversions from the Cleminson composites of 1882, having a single saloon, and were third class only. The oldest, however, No. 2, was built in 1875 as a six-wheeled composite and was 32 ft long with an entrance vestibule at one end, a single third class saloon, and a guard's compartment at the other end. In 1899 one bogie first class saloon was built, 42 ft long, with guard's compartment at one end, two side-by-side toilets, and two small saloons (No. 1). In 1891, No. 3 appeared, another six-wheeled first class saloon, similar to No. 1 in arrangement but on a 34 ft 6 in. under-frame. Finally, in 1900-1, two bogie third class saloons with luggage compartment were brought out, on 42 ft frames, Nos. 6 and 7. Though bogie vehicles were being built from 1897, odd six-wheelers were still being turned out contemporaneously, the last being the two semi-corridor composites of 1900-1. These, as with other six-wheelers built from 1895 onwards, owed a great deal to similar vehicles of Aston's design for the Cambrian Railways – no doubt the coming of Mr Aslett to the Furness Railway from the fastnesses of mid-Wales had some bearing on this. It was Aslett who really put the Furness passenger services on their feet (or should one say wheels?). In 1894 it was forced upon the company to purchase 55 six-wheelers fitted with continuous automatic vacuum brakes, through the edict of the Board of Trade, and delivery of these was taken in 1895-96. Most were five-compartment thirds, but a few brake thirds were included in the contract, which was placed with the Gloucester Railway Carriage & Wagon Co. Six five-compartment thirds were built by Ashbury's of Manchester in 1897.

Another notable point with the FR methods of passenger operation was a predilection for the use of a full brake van on secondary and branch passenger trains, in place of a brake third or brake composite coach. The company had a number of these, both four- and six-wheeled, and numbered in a separate list for non-passenger carrying stock. Thus the proportion of brake thirds and brake composites to ordinary stock was lower than normal, a feature which applied to several of the smaller railway companies (notably in South Wales) who favoured this mode of operation. The workmen's trains in West Cumberland were usually made up of three or four four-wheelers and a full brake, though one or two of the older six-wheelers were sometimes used. Since there were only fourteen corridor vehicles all told, a complete train of such stock was hardly, if ever, seen; they were employed mainly as through coaches to distant termini over other companies' lines. Though the total distance from Carnforth to Whitehaven was approximately 75 miles, the ordinary non-corridors and semi-corridor stock could cope quite adequately. Such complete corridor trains as were seen on the line were mainly LNWR or Midland stock, working to Barrow from London, Birmingham, or Leeds.

The LMS renumbered the whole of its passenger stock in 1933-34 to bring stock of various origins into blocks of numbers according to their type and length. The original 1923 renumbering was by blocks of numbers allocated to each constituent company, regardless of type, though there was some system of sorting out within each block,

Six-wheeled carriage. *John Alsop collection*

particularly regarding the larger companies. The Furness stock was renumbered with no easily recognisable pattern; apart from the fact that the saloons became 15001-15007, there does not seem to be any real system with the remainder, though to some extent coaches on the same Furness diagram came into the same block of numbers, but there were exceptions even to this. The FR. block was 15001-15300. However, by comparison of the 1923 and 1933 renumbering, the surprising fact is elicited that the whole of the 112 FR bogie carriages included in the 1923 scheme were still in service in 1933 and can be cross-referenced (their original FR numbers are unfortunately lost), a state of affairs which does not apply to most of the other constituent companies. Only about 60 of the six-wheelers had been withdrawn during these ten years, in contrast to the L&YR, for example, whose six-wheelers ran into several hundreds in 1923 but were practically non-existent in 1933. The few surviving Furness four-wheelers, all classified as "workmen's coaches", were soon withdrawn. The exact number is not known, but by computation would seem to be about 25 in all, most of which were on the "A" list. This total does not, however, include the two rail-motor trailers built in 1905-7 to work with the steam rail motors. Each was a composite four-wheeled saloon (see p. 141) and had a luggage compartment at the rear, in which the control gear for reverse running was fitted. Like the rail motors. themselves, they had clerestory roofs, the only vehicles to be built to this pattern. Numbered 123 and 193, they continued in use as ordinary coaches for some time after the withdrawal of the rail motors in 1914, but do not appear to have survived the Grouping.

There were also a number of non-passenger vehicles of various types which, through being fitted with continuous brakes, were classified for running with passenger trains. These included two Post Office sorting vans, Nos. 1 and 2, which were vastly different from each other, though both had six wheels. No. 1, built in 1887, was a very short van, only 22 ft 6 in. long, while No. 2, built in 1903, was ten feet longer. Both were similar in internal arrangements, though very different externally. Neither of them had corridor connections, since they were designed not to work together, and they had no pick-up net, as the Furness never had any lineside apparatus. The 24 full brake vans have been mentioned; besides these were 22 horse boxes, in two sizes, nine open carriage trucks, four ventilated vans, three closed carriage vans, and one prize cattle truck, the whole of this stock having been built between 1883 and 1908.

A number of coaches, and other stock, were fitted with dual brakes- vacuum and Westinghouse for working over North Eastern Railway lines; some had complete dual brakes, while others had only through pipes for the Westinghouse system. Furness passenger stock was painted dark blue below the waist panels and white above, the end panelling being blue all over. All mouldings were also blue, and lining was in white and yellow. Class lettering on door waist panels, the initials FR, and numbers were gold, outlined dark blue. Grab handles were of a peculiar shape, rather like a right-angled letter Z. Under-frames and running gear were black and roofs grey. Some wheel tyres were painted white, but this was not universal. A peculiar point was that torpedo ventilators were only fitted to smoking compartments, of which there were never more than two in the largest coaches; brake thirds and six-wheelers had only one, usually at one end. Such compartments were clearly marked on the door waist panel "THIRD SMOKING" in two lines. There do not appear to have been any first class smoking compartments; clearly the Furness Railway regarded smoking as the prerogative of the hoi polloi. In later years, begun mainly as a war economy measure, carriages became blue all over, the roofs, however, remaining grey. Some vehicles such as horse boxes, etc., did not have any white panelling. Furness passenger stock was not luxurious but was quite adequate and comfortable, comparing favourably with many other companies' stock. Earlier coaches were all obtained from outside contractors, but from 1882 onwards a number of coaches were built in the company's own shops at Barrow, which were well equipped for all kinds of repair work, both for passenger and goods vehicles.

The Furness Railway produced a *Diagram Book* for passenger stock (also one for goods) but there is considerable doubt as to the date. Probably it was drawn up specially for the Grouping, when all constituent companies were asked to send diagrams of their rolling stock to Derby. At all events, the *Diagram Book* was not in existence much before 1923. There was apparently no chronological order, since old and new vehicles were mixed together; the only attempt at separation was by classes. Thirds were on diagrams 1-16, composites 17-26, and saloons 27-31. Other diagrams up to 42 covered full brakes, carriage trucks, and so on. No diagram was issued for the two rail-motor trailers, which seems to indicate that they had been withdrawn by 1923. The only four-wheeled coaches in the book were four-compartment thirds on diagram 8 and two-compartment brake thirds on diagram 15, and in neither case was the total number stated, as was the case with all other diagrams. Those remaining in 1923 were all in West Cumberland for workmen's trains and were not included in the 1923 LMS renumbering, unless they fitted into some of the otherwise inexplicable blank numbers in the series. No complete Furness Railway passenger stock list exists, at least, not from official sources, and only odd original numbers are known, mainly from photographs. The official total of FR coaches in 1923 was 281 (plus four built in 1924) for which the LMS had allowed a block of 300 numbers, though it is by no means certain that any numbers above 15285 were ever allotted. Since the highest original Furness number was 336, the list must have contained some 50 blanks, and as in addition the four-wheelers were mostly on the "A" list, the position is extremely complicated.

Goods Stock

As might be expected from the main type of traffic on the system, the bulk of the goods stock consisted of open wagons of various designs and sizes, quite a large proportion of them being specially constructed for the conveyance of iron-ore, limestone, and coke. Covered vans were a comparatively small part of the stock, numbering about 300 out of the total stock of 7,428. As the highest number carried was 7,880, there were thus a considerable number of blanks. The open wagons varied from 6-ton one-plank types to 20-ton high-sided coal wagons, all four-wheeled, of which 1,167 were classed as ore wagons, most of them having hopper doors in the floor. In addition to these there were 99 steel hopper wagons of 12 tons capacity, built between 1907 and 1915. Some large open coke

A goods train at Grange-over-Sands with a train of empty mineral wagons.
Oakwood Press

wagons, 60 in all, were built in 1905, in two sizes, 18 ft and 20 ft long, but both rated at 10 tons. There were very few bogie wagons, only seven in all, of which one, 7536, was constructed at Barrow specially for the conveyance of large naval guns from Vickers' works. This was 44 ft long over headstocks, with a 26 ft well, and ran on plate-framed bogies of only 4 ft wheelbase, fitted at 35 ft centres, and it was rated at 35 tons. Six bogie bolster wagons of 25 tons capacity were built between 1918 and 1924, having 5 ft bogies at 24 ft centres, and were 34 ft long. Other wagons for special traffic were 25 wood pulp wagons, 23 ft long, and rated at 10 tons, running on four wheels, and 50 similar rail wagons of 15 tons capacity, on six wheels. Six steel-bodied explosives wagons were provided for traffic to the mines in the area, and the Irish cattle traffic was catered for by 132 cattle wagons of various patterns, some without roofs. There were 87 goods brake vans, of three main types, all rated at 10 tons, and all having an open verandah at both ends. Mention should be made of a batch of "white elephants', in the shape of about 20 enormous bogie wagons specially built for iron-ore traffic, and with a rated capacity of 45 tons. These were entirely of wood and were built in 1913, but in spite of their size they tared only 12¾ tons. Unfortunately, they proved very uneconomical in use and were extremely difficult to unload in contrast to the steel hopper wagons which could discharge their full load in less than a minute. So by the end of 1920 the bogie ore wagons had disappeared. Their ultimate fate is not known, but it is believed that they were broken up. They caused trouble also in sidings, on account of their length, and though a bold experiment, no one was sorry to see the last of them.

A large number of wagons were built in the Furness Railway shops at Barrow, but the bulk were obtained from outside contractors. Livery was light grey, with white lettering and black ironwork. The initials FR were 17½ in. high and numbers, in 6 in. figures, were painted on the bottom plank at the left-hand end. Numbers were also painted on wagon ends, in 3 in. figures, in the left-hand bottom comer.

Barrow Works were well equipped and covered an area of 30 acres. The original fitting shop opened in 1846 for the repair and maintenance of locomotives, and eventually became the machine and turning shop for the works. The new erecting shop had three bays, each of 50 ft span and 480 ft long, and as first built had a low roof, so engines had to be moved outside for the removal of wheels. About 1880 the roof was raised and an overhead travelling crane erected, with a capacity of 30 tons. The adjoining boiler shop was equipped with pneumatic riveters, sheet bending machines, and a forge, and also had a 15-ton overhead crane. Both this, and the erecting shop crane, were converted to electric operation in 1908.

As well as the locomotive shops, there were large carriage and wagon building bays, the latter being 300 ft. long and 160 ft wide, and the carriage shops 210 ft long by 120 ft wide. There was a separate paint shop 210 ft in length, with a capacity of 20 coaches. There was also a wagon repair shop and a large timber drying shed. Though a large number of wagons and several coaches were built in the works, rather surprisingly only two locomotives were ever built, these being the two rail motors. All heavy overhauls and general repairs were dealt with very successfully. The works also carried out engineering and maintenance for the dock equipment.

Near the works was the main locomotive shed, 310 ft long and 150 ft wide, with an allocation of 60 locomotives. Four carriage storage sheds near Central station had a capacity of 152 coaches, more than half the total stock. There were two other major locomotive sheds, Carnforth and Moor Row, both capable of carrying out heavy repairs. Other sheds, with small locomotive allocations, were at Whitehaven, Lakeside, Coniston, Workington Central, and Siddick Junction. One or two engines were also stabled overnight in the LNWR sheds at Oxenholme and Lancaster.

Like many other railways the Furness Railway held a shunting competition every year. The object being to couple and uncouple a line of wagons as quickly as possible. The first competition was held in 1887 and they continued until the Grouping.

John Alsop collection

Locomotive improvement class, 1912. *Oakwood Press*

Ulverston station staff. *John Alsop collection*

Chapter Eleven

Accidents

The Furness Railway was one of those very few companies in Britain which could boast of a very good record, as far as safety of passengers is concerned. However, there were one or two accidents of note, whereby railway employees were killed, and one accident in particular in which a relatively large number of passengers were injured. However, this latter accident, as will be seen, was not of the company's causing.

Preston Street Station, Whitehaven

Before Preston Street was closed to passengers, a very haphazard method of dealing with incoming trains was in force. The station was at the foot of a slight falling gradient. All trains were stopped 500 yards from the platform, the engine was uncoupled and run forward into a siding. The only brake power then available on the train were the four wooden brake blocks on the guard's van. The guard would release his brake and the carriages were allowed to run by gravity into the platform, after which the engine ran out from the siding and coupled onto the other end of the train. On several occasions, the application of the guard's hand brake came too late to prevent a collision with the buffer-stops, several times with more than usual force, thereby giving passengers a severe shaking. There was only one serious accident, however, due to complete failure of the guard's brake. After this accident the practice was abandoned. Let it be said that this practice was by no means uncommon all over the country, and on a number of occasions (such as at Burnley on the L&YR in 1852) led to considerable loss of life. The Whitehaven examples, happily, did not have any fatal results.

Broughton-in-Furness, 1861

Before the amalgamation of the Furness and Whitehaven & Furness Junction Railways had taken place, the two companies managed to stage a spectacular accident together at Broughton.

The Furness Railway's 2-2-2 well-tank No. 12, newly built, was working a passenger train from Foxfield to Coniston. It was the custom that vans or wagons for Broughton, or for exchange with the

Whitehaven company, were pushed in front of the engine, since the entrance to Broughton yard was on a junction facing Foxfield. Meanwhile, the Whitehaven company's 2-2-2 tank *Oberon* had left Broughton, light, to pick up a train for the north. Her driver and fireman saw No. 12 approaching but realised that the crew of the Furness engine could not see them because of the van which was being propelled in front of the train. *Oberon* was immediately thrown into reverse gear, but this was not sufficient to prevent a collision. The fireman of the Furness engine was riding on the front buffer beam ready to detach the van at the yard points, and when the collision occurred he was immediately crushed by the wreckage of the van. The crew of the *Oberon* jumped off at the last moment, but were unhurt, and were chagrined to find that their engine, having responded to her reverse gear, was running away up the line towards Coniston. Fortunately, the gatekeeper at Broughton saw her coming and opened the gates, the engine finally coming to rest, short of steam, on the 1 in 49 gradient near Torver.

Bransty Tunnel, Whitehaven, 1866

The single-line tunnel between Corkickle and Bransty stations at Whitehaven was the scene of an accident in 1866. All trains through the tunnel were worked by a special pilot engine, without which trains were not supposed to proceed. A pilot man was employed at Corkickle to give right of way to trains entering the tunnel. Be it said that the telegraph wire through the tunnel had long been inoperative. However, the notable safety precaution of the pilot was not always adhered to; when two trains were to proceed in the same direction without an intervening train the opposite way, it was customary to let the second train proceed after a 15-minute interval. On this occasion, a mineral train for Maryport was sent off through the tunnel with the pilot engine in charge. En route through the tunnel a coupling broke and left eight wagons and the guard's van standing on the line. The driver, not having felt the "rug" on his engine, proceeded to Bransty, unaware that anything was amiss. Before anything could be done about the situation, the pilot man, after the 15-minute interval, gave right of way to a passenger train again hauled by *Oberon* for Bransty. In the tunnel this train collided violently with the standing wagons, the driver of *Oberon* being badly scalded and the fireman killed outright. It was thought, but never definitely established, that as at Broughton the fireman had been at the front of the engine putting sand on the rails, as there was no automatic sanding gear, and the continual dampness of the tunnel made this method customary. Several passengers complained of minor injuries, but the train was travelling at a low speed, otherwise the casualty list might have been much greater.

Ravenglass, 1866

In the same year as the tunnel accident, a further accident occurred on the W&FJR line at Ravenglass. The 0-6-0 tender engine *Lonsdale* had been shunting in the sidings, and after detaching a number of wagons resumed its journey towards Whitehaven. However, the points for the siding had been left open, and the train ran over them before the crew realised what had happened. After demolishing the stop blocks, the engine and a wagon containing 20 sheep went over the end but did not overturn. All the sheep escaped unhurt, but the driver, fireman, and guard were on the footplate, and were badly scalded by steam escaping from a fractured pipe. The engine was otherwise undamaged and was hauled back onto the siding by means of a temporary length of track.

Lindal Sidings, 1892

One of the most peculiar accidents ever to happen to a train took place in Lindal sidings during the morning of 22nd October, 1892. A standard Sharp 0-6-0, No. 115, was shunting in the ore sidings, and in the course of its movements came out onto the main line. Almost immediately the ground gave way beneath the engine, which slipped into the hole at an angle, supported by the front of the smokebox. The crew jumped clear and were unhurt. Within minutes the ground had subsided still further, and by 3.00 pm the engine had gradually disappeared into the hole. The coupling between engine and tender broke, and the latter, though derailed, was drawn back with the rest of the train to safety. By the time that the subsidence had stabilised itself it had involved eight lines of railway, and all traffic on the main lines had been stopped. Work began immediately in filling the hole, but it was not until the following spring that normal traffic was resumed. The engine was completely buried and has lain there ever since.

For several weeks passenger trains were brought up to either side of the crater and passengers walked round to join another train waiting at the opposite side. Later it was found possible to run trains through, at a

The crater formed by the Lindal Siding subsidence. Locomotive No. 115 is still lying in the old mine workings, though it is unlikely that there would be anything salvageable today, from the damp and corrosive conditions of its tomb. *John Alsop collection*

walking pace, over a distant siding which had only been slightly affected, but even so, it was thought wisest to detrain the passengers while this manoeuvre was carried out. Goods traffic was seriously delayed, for it had to be sent round by an extremely circuitous route through Tebay, Penrith, Keswick, Workington, and Millom.

The Lindal area is honeycombed with old iron-ore workings, and it was thought that underground streams, seeping into the workings, caused the subsidence and consequent loss of the engine, which is computed to be nearly 200 feet underground.

Dalton Station, 1898

During the day there was only one passenger train along the loop line from Askam to Dalton, this being provided for the benefit of businessmen, and worked only in one direction in the morning. The train usually comprised a composite brake coach and a normal third class six-wheeler, though it was occasionally strengthened by the addition of a second six-wheeler. After the passengers had detrained, the coaches were shunted into the Lindal bay at Dalton station, and remained there until the afternoon, when they were worked back over the loop line. On 16th June, 1898, while the carriages were being shunted, they broke loose from the engine, ran back, and caused considerable damage to the platform, and themselves. The guard had stayed on the train while it was being drawn up the incline towards the tunnel; on the gradient the breakaway occurred, and the signalman was foresighted enough to switch the points rapidly for the bay. The guard found the braking power of his train insufficient, but stuck to his post, and was seriously injured in the collision. The signalman was awarded an honorarium of 20 shillings and a letter of commendation for his action in preventing a more serious accident, for if he had not switched the bay points the carriages would have run back through the station and on a continually falling grade, would have been derailed at the junction, probably blocking all lines.

Leven Viaduct, 1903

The night of 26th-27th February, 1903, was one of serious storms. It left a trail of disaster from the mid-Atlantic to the Shetlands. The Post Office officials were making preparations at Ulverston for the arrival of the mail train from the South. It was noted that the train was long overdue, and a telephone call to the station elicited the fact that all communication with Carnforth was broken and the position of the train was not known. A later phone call, however, established the fact that a gust of wind of over 130 mph had blown the train off the rails on the Leven viaduct. Helpers were gathered from all parts of the town, and the party set out in the teeth of a howling gale to the scene of the accident. In the meantime, a goods train had arrived from Barrow, and after a van had been obtained from a siding, the goods engine set out with the van, packed with helpers, to see what could be done. Also present were a number of railway officials who had been notified of the accident. To approach closely with the engine and van was out of the question, so the train halted at the end of the viaduct and the rescuers set out on foot to the scene in single file and keeping a firm grip on the handrails. How nearly a second Tay Bridge Disaster had been averted could only be realised by those present, for the train lay completely on its side, with the exception of the locomotive, which stood firmly on the rails, and the mail van, coupled next to the tender, which was oscillating dangerously on one wheel. This was the new Carnforth and Whitehaven sorting van; the train consisted of ten vehicles, the sorting van, six parcel vans, and three passenger coaches. Fortunately the train was lying on

Dalton station 11th August, 1913 where an engine shunted these coaches into the stop block and onto the platform. No one was on board at the time and nobody was hurt, not even the boy with his arm in a sling, whom the photographer has posed for dramatic effect. A breakdown team eased the coaches back on the railway later that day.

John Alsop collection

This accident occurred north of Egremont at Gillfoot Junction on 15th September, 1903. A rail broke as the locomotive, at the head of 18 wagons of iron-ore, passed over, derailing it and causing the train to come to a very rapid halt. The weight of the wagons slammed into the back of the locomotive crushing the coalbox which fatally injured one of the men on board.

John Alsop collection

the down track, the wind coming from the south-west. Had the train been going in the other direction- it would have been most certainly blown into the estuary. Even so, the last coach was hanging partly over the side of the viaduct. The strength of the wind may be judged by the fact that the tide was being held back in the estuary 12 to 15 feet above high water level, and the surface of the water was less than two feet below the bridge decking.

A company official ascertained that immediately after the accident, in great danger to themselves, the railway employees on the train had proved themselves heroes. After struggling along the line of overturned carriages they had dragged passengers from the train through broken windows and formed a human ring around them, and in this manner had succeeded in getting them to the shelter of cottages at the Ulverston end of the viaduct.

The train had had a most dangerous trip; several times broken telegraph wires had tangled themselves round wheels, and had even caused the severance of a vacuum brake pipe, but after a consultation at Grange, where it was suggested that the train be terminated and the mails and passengers sent on by road, the crew decided to stick it out and try to get through. Early next morning, the breakdown crane, with three or four spare engines, arrived from Barrow, with Mr Rutherford himself in charge, as it had been decided the wisest course was not to begin clearing-up operations until the storm had abated, otherwise there was the probability of further disaster in such an exposed position.

The Leven viaduct in the aftermath of the accident in 1903. *Oakwood Press*

The difficulties of the breakdown gang can best be imagined when it is remembered that the train was lying on its side athwart the down track, with the last coach slewed across and effectively blocking both lines. The first operation was to take a firm grip on the oscillating sorting van; it was fixed in a sling and raised, when it began to swing about like a gigantic pendulum with the force of the wind. However, it was replaced on the track after a terrible struggle, and was taken off by one of the spare engines to deliver its mail at its accustomed stops along the line to Whitehaven. Near Nethertown, the Glasgow vessel *Viscount* was driven ashore close to the track, her captain having been lost overboard in the storm. The mail van arrived at Whitehaven at 2.00 pm on 27th February.

Because of the awkward position of the train on the viaduct, an additional engine was brought up from Carnforth, and with a standard Sharp 0-6-0 at each end, the train was slowly pulled and pushed, still on its side, to the Ulverston end of the viaduct, where the breakdown crane righted the vehicles and replaced them on the rails. The nine coaches were then run together over Lindal to Barrow Works, where they were repaired and returned to service. In this accident, 33 passengers were injured in varying degrees, mainly by broken glass and bruises in being thrown about when the train overturned. None was seriously hurt.

Arnside, 1917

During 1916 and 1917, Arnside Viaduct was being reconstructed, and there were special points laid in at each end so that one half of the viaduct could be dealt with while trains ran on a single track on the other half. The winter, however, was very severe, and one morning, after a heavy frost, a long goods train, hauled by a Pettigrew 0-6-0 and a 0-6-2 tank, while negotiating the bridge, was derailed by the points being frozen solid at the Arnside station end. The locomotives were not damaged, but wholesale destruction was caused amongst the wagons, which crashed over the side of the embankment. There were no injuries in this accident, although it was necessary to close the line for 48 hours to allow clearing-up operations to proceed.

Appendix One

Locomotive List

BUILDERS

BCK	Bury, Curtis & Kennedy, Liverpool.	NBL	North British Locomotive Co., Glasgow.
AB	Andrew Barclay, Sons & Co., Kilmarnock.	NW	Nasmyth, Wilson & Co., Manchester.
EBW	E. B. Wilson, Leeds.	RS	Robert Stephenson & Co., Darlington.
F	William Fairbairn & Co., Manchester.	SL	Stothert & Slaughter, Bristol.
FJ	Fletcher, Jennings & Co., Whitehaven.	SB	Sharp Brothers, Manchester.
		SS	Sharp, Stewart & Co., Manchester.
H	R. & W. Hawthorn, Newcastle-on-Tyne.	VF	Vulcan Foundry, Newton-le-Willows.
K	Kitson & Co., Leeds.	R/00	denotes rebuilding date, last two figures being the year.
N	Neilson & Co., Glasgow.		

In connection with this, T2, T3, T 4 denote the type of boiler fitted at rebuilding to the D1 class 0-6-0's, and LY denotes fitted with L. & Y.R. saturated Belpaire boilers at Horwich Works, (D3 and D4 0-6-0's; L2 and L3 0-6-2 tanks.). Renumberings are shown by the new number followed by the year in brackets, thus-32A (1896).

No.	Class	Builders	Date	Withdrawn	Notes
1	A1	BCK	1844	1867	
2	–"–	BCK	–"–	1871	Sold
3	A2	BCK	1846	1900	Preserved
4	–"–	BCK	–"–	1898	
5	B1	SB 696	1852	1873	
6	–"–	SB 697	–"–	1873	
7	A3	F	1854	1899	
8	–"–	F	–"–	1899	
9	–"–	F	1855	1901	9A (1899)
10	–"–	F	–"–	1900	10A (1899)
11	B2	SB 1016	1857	1873	Sold
12	–"–	SB 1017	–"–	1898	Sold, 12A (1873)
13	A4	F	1858	1900	13A (1899)
14	–"–	F	–"–	1900	14A (1899)
15	–"–	F	1861	1899	
16	–"–	F	–"–	1899	16A (1899)
17	A5	SS 1434	1863	1870	Sold
18	–"–	SS 1435	–"–	1870	Sold
19	–"–	SS 1447	–"–	1870	Sold
20	–"–	SS 1448	–"–	1870	Sold
21	B3	SS 1500	1864	1898	21A (1896)
22	–"–	SS 1501	–"–	1898	Sold, 22A (1896)
23	C1	SS 1543	–"–	1904	98 (1898)
24	–"–	SS 1544	–"–	1904	99 (1898)
25	A5	SS 1585	1865	1873	Sold

No.	Class	Builders	Date		Withdrawn	Notes
26	A5	SS 1586	1866		1873	Sold
27	–"–	SS 1663	–"–		1918	27A (1914)
28	–"–	SS 1664	–"–		1918	28A (1914)
29	D1	SS 1697	–"–		1925	R/00 T2, 61 (1918), LMS 12001
30	–"–	SS 1698	–"–		1925	R/99 T2, 62 (1918), LMS 12007
31	–"–	SS 1764	–"–		1925	R/99 T2, 48 (1920), LMS 12000
32	–"–	SS 1765	–"–		1900	32A (1896)
33	–"–	SS 1766	–"–		1900	33A (1896)
34	B3	SS 1763	–"–		1898	34A (1896)
35	–"–	SS 1768	–"–		1898	Sold 35A (1896)
36	–"–	SS 1767	–"–		1898	36A (1896)
37	–"–	SS 1762	–"–		1898	37A (1896)
38	D1	SS 1760	–"–		1915	
39	–"–	SS 1761	–"–		1915	
40	–"–	SS 1784	–"–		1925	R/00 T2, 80 (1916), LMS 12008
41	–"–	SS 1785	–"–		1916	
42	D2	H 1269	1864	Ex-WFJ 19	1904	R/86
43	–"–	H 1245	–"–	Ex-WFJ 18	1873	Sold
44	F1	H 997	1857	Ex-WFJ 3	1882	
45	–"–	H 998	–"–	Ex-WFJ 13	1882	
46	B4	H 1148	1860	Ex-WFJ 10	1876	Sold
47	B5	EBW	1850	Ex-WFJ 4	1870	Sold
48	–"–	–"–	–"–	Ex-WFJ 5	1870	Sold
49	C2	FJ 29	1862	Ex-WFJ 15	1882	Sold
50	–"–	N 571	1863	Ex-WFJ 16	1882	Sold
51	G1	SS 1842	1867		1915	

G1 Locomotive No. 51.   *John Alsop collection*

LOCOMOTIVE LIST

No.	Class	Builders	Date	Withdrawn	Notes
52	G1	SS 1843	1867	1918	84 (1915)
1	E1	SS 2057	1870	1916	1A(1913)
2	-"-	SS 2058	-"-	1918	2A (1913)
17	D1	SS 2064	1871	1925	R/00 T2, 23 (1900), 42 (1910), 66 (1916) LMS 12003
18	-"-	SS 2065	-"-	1910	24 (1900)
19	-"-	SS 2095	-"-	1910	
20	-"-	SS 2096	-"-	1925	R/00 T2, 25 (1910), 25A (1913), LMS 12002
53	-"-	SS 2097	-"-	1916	
54	-"-	SS 2098	-"-	1925	R/98 T2, 78(1916), LMS 12005
55	-"-	SS 2099	-"-	1918	
56	-"-	SS 2100	-"-	1913	
57	E1	SS 2093	-"-	1918	
58	-"-	SS 2094	-"-	1918	
59	-"-	SS 2145	-"-	1913	
60	-"-	SS 2146	-"-	1921	R/00 T2, 64 (1918)
61	-"-	SS 2147	-"-	1916	R/00 T2
62	-"-	SS 2148	-"-	1916	R/99 T2
63	-"-	SS 2149	-"-	1918	R/00 T2
64	-"-	SS 2150	-"-	1918	
65	-"-	SS 2151	-"-	1930	R/18 T4, LMS 12065
66	-"-	SS 2152	-"-	1916	
67	-"-	SS 2153	-"-	1914	
43	-"-	SS 2154	-"-	1925	R/01 T2, 67(1916), LMS 12004
68	G1	SS 2204	1872	1925	LMS 11549
69	-"-	SS 2205	-"-	1925	LMS 11550
70	E1	SS 2245	-"-	1924	R/91 (2-4-2T), 70A (1920), LMS 10619
71	-"-	SS 2246	-"-	1923	R/91 (2-4-2T), 71A (1920), LMS 10620
72	-"-	SS 2247	-"-	1919	R/91 (2-4-2T)
73	-"-	SS 2248	-"-	1919	R/91 (2-4-2T)
74	-"-	SS 2249	-"-	1921	R/91 (2-4-2T), 72A (1920)
75	-"-	SS 2257	-"-	1914	
46	-"-	SS 2256	-"-	1920	
47	-"-	SS 2258	-"-	1919	R/91 (2-4-2T)
48	-"-	SS 2259	-"-	1920	R/91 (2-4-2T)
25	D1	SS 2278	1873	1910	
26	-"-	SS 2279	-"-	1930	R/16 T4, 59 (1913), 63 (1918), LMS 12066
76	-"-	SS 2280	-"-	1925	R/01 T2, LMS 12006
77	-"-	SS 2283	-"-	1914	
78	-"-	SS 2284	-"-	1915	R/99 T2
79	-"-	SS 2285	-"-	1930	R/16 T4, LMS 12067
80	-"-	SS 2316	-"-	1916	
81	-"-	SS 2317	-"-	1921	R/00 T2
82	G1	SS 2300	-"-	1925	LMS 11551
83	-"-	SS 2301	-"-	1925	LMS 11552
84	D1	SS 2337	-"-	1915	
85	-"-	SS 2338	-"-	1925	R/00 T2, LMS 12009

No.	Class	Builders	Date	Withdrawn		Notes
86	D1	SS 2340	1873		1925	R/98, T2 LMS 12010
87	–"–	SS 2341	–"–		1924	R/98, T2 LMS 1201 l
5	E1	SS 2364	–"–		1907	
6	–"–	SS 2365	–"–		1907	
11	–"–	SS 2366	–"–		1916	3 (1899), 3A (1907)
12	–"–	SS 2367	–"–		1920	4 (1899), 4A (1907)
92	D1	SS 2422	1874		1924	R/97 T2, 75 (1914), LMS 12012
93	–"–	SS 2423	–"–		1925	R/99 T2, 77 (1914), LMS 12013
94	C1	SS 2448	–"–		1914	94A (1912)
95	–"–	SS 2449	–"–		1916	95A (1912)
96	–"–	SS 2450	–"–		1916	96A (1907)
97	–"–	SS 2451	–"–		1924	97A (1907), LMS 11258
88	D1	SS 2506	1875		1926	R/11 T3, LMS 12068
89	–"–	SS 2507	–"–		1926	R/12 T3, LMS 12069
90	–"–	SS 2508	–"–		1930	R/18 T4, LMS 12070
91	–"–	SS 2509	–"–		1924	R/00 T2, Dept. No. 1 (1918)
98	G2	RS 1008	1855	Ex-WCE 1	1895	
99	–"–	RS 1009	–"–	Ex-WCE 2	1886	
100	–"–	FJ 21	1858	Ex-WCE 4	1918	R/95, 100A (1904)
101	–"–	RS 1310	1860	Ex-WCE 5	1890	
102	–"–	RS 1437	1862	Ex-WCE 6	1900	
103	–"–	RS 1487	1863	Ex-WCE 7	1887	
104	–"–	RS 1488	–"–	Ex-WCE 8	1914	104A (1904)
105	–"–	RS 1798	1867	Ex-WCE 9	1921	105A (1904)
106	–"–	RS 1804	1869	Ex-WCE 10	1918	106A (1904)
107	–"–	RS 1960	1870	Ex-WCE 11	1918	107A (1904)
108	H1	SL	1850	Ex-WCE 12	1898	
109	G2	RS 1997	1871	Ex-WCE 13	1925	109A (1907), LMS 11547
110	–"–	RS 2109	1873	Ex-WCE 14	1920	110A (1907)
111	–"–	RS 2110	–"–	Ex-WCE 15	1920	111A (1907)
112	G3	AB 154	1875	Ex-WCE 17	1925	108 (1904), 108A (1907), LMS 11548
113	G4	H 989	1857	Ex-WCE 3	1898	
114	D1	SS 2945	1881		1926	R/10 T3, 115 (1898), 70 (1920), LMS 12071
115	–"–	SS 2946	–"–		1892	(Lindal subsidence)
116	–"–	SS 2947	–"–		1925	R/01 T2, 71 (1920), LMS 12014
117	–"–	SS 2948	–"–		1927	R/13 T3, 72 (1920), LMS 12072
118	–"–	SS 2949	1881		1927	R/11 T3, 73 (1920), LMS 12073
119	–"–	SS 2950	–"–		1926	R/10 T3. 74 (1920) LMS 12074
44	E1	SS 3086	1882		1925	R/98, 44A (1920), LMS 10002
45	–"–	SS 3087	–"–		1921	R/91, 45A (1920)
49	D1	SS 3170	1883		1928	R/16 T4, LMS 12075
50	–"–	SS 3171	–"–		1928	R/12 T3, LMS 12076
120	–"–	SS 3172	–"–		1887	Sold
121	–"–	SS 3173	–"–		1887	Sold
120	K1	SS 3618	1890		1927	LMS 10131
121	–"–	SS 3619	–"–		1928	LMS 10132
122	–"–	SS 3620	–"–		1927	LMS 10133

LOCOMOTIVE LIST

No.	Class	Builders	Date	Withdrawn	Notes
123	K1	SS 3621	1890	1925	LMS 10134
21	K2	SS 4174	1896	1930	32 (1896), 44 (1910), LMS 10137
22	-"-	SS 4175	-"-	1928	33 (189,6), 45 (1910), LMS 10138
34	-"-	SS 4176	-"-	1927	46 (1920), LMS 10139
35	-"-	SS 4177	-"-	1928	47 (1920), LMS 10140
36	-"-	SS 4178	-"-	1929	LMS 10135
37	-"-	SS 4179	-"-	1931	LMS 10136
112	L1	SS 4364	1898	1927	LMS 11622
113	-"-	SS 4365	-"-	1928	LMS 11623
114	-"-	SS 4366	-"-	1928	LMS 11624
7	D3	NW 552	1899	1928	LMS 12468
8	-"-	NW 553	-"-	1930	LMS 12469, R/25 LY
9	-"-	NW 554	-"-	1932	LMS 12470
10	-"-	NW 555	-"-	1930	LMS 12471
11	-"-	NW 556	-"-	1930	LMS 12472
12	-"-	NW 557	-"-	1929	LMS 12473
13	-"-	SS 4563	-"-	1930	LMS 12474
14	-"-	SS 4564	-"-	1932	LMS 12475
15	-"-	SS 4565	-"-	1930	LMS 12476
16	-"-	SS 4566	-"-	1930	LMS 12477
17	-"-	SS 4567	-"-	1929	LMS 12478
18	-"-	SS 4568	-"-	1936	LMS 12479, R/25 LY
124	K2	SS 4651	1900	1928	LMS 10141
125	-"-	SS 4652	-"-	1929	LMS 10142
126	K3	SS 4716	1901	1931	LMS 10143

K3 locomotive No. 126. *John Alsop collection*

No.	Class	Builders	Date	Withdrawn	Notes
127	K3	SS 4717	1901	1930	LMS 10144
128	–"–	SS 4718	–"–	1930	LMS 10145
129	–"–	SS 4719	–"–	1930	LMS 10146
98	L2	NW 689	1904	1935	LMS 11625
99	–"–	NW 690	–"–	1930	LMS 11626
100	–"–	NW 691	–"–	1936	LMS 11627
101	–"–	NW 692	–"–	1946	LMS 11628
102	–"–	NW 693	–"–	1930	LMS 11629
103	–"–	NBL 16113	–"–	1934	LMS 11630, R/27 LY
104	–"–	NBL 16114	–"–	1930	LMS 11631
105	–"–	NBL 16115	–"–	1933	LMS 11632
106	–"–	NBL 16116	–"–	1931	LMS 11633, R/27 LY
107	–"–	NBL 16117	–"–	1929	LMS 11634
3	D4	NBL 17840	1907	1930	LMS 12480, R/26 LY
4	–"–	NBL 17841	–"–	1930	LMS 12481
5	–"–	NBL 17842	1907	1934	LMS 12482, R/26 LY
6	–"–	NBL 17843	1907	1930	LMS 12483, R/26 LY
96	L3	NBL 17808	–"–	1938	LMS 11635, R/27 LY
97	–"–	NBL 17809	–"–	1941	LMS 11636, R/27 LY
108	–"–	NBL 17810	–"–	1935	LMS 11637
109	–"–	NBL 17811	–"–	1933	LMS 11638
110	–"–	NBL 17812	–"–	1931	LMS 11639
111	–"–	NBL17813	–"–	1933	LMS 11640
19	G5	VF 2523	1910	1942	55 (1918), LMS 11553
20	–"–	VF 2524	–"–	1930	56 (1918), LMS 11554

G5 locomotive No. 24. *John Alsop collection*

LOCOMOTIVE LIST

No.	Class	Builders	Date	Withdrawn	Notes
21	G5	VF 2525	1910	1930	57 (1918), LMS 11555
22	–"–	VF 2526	–"–	1932	58 (1918), LMS 11556
23	–"–	VF 2527	–"–	1932	59 (1918), LMS 11557
24	–"–	VF 2528	–"–	1935	60 (1918), LMS 11558
94	L4	K 4855	1912	1934	LMS 11641
95	–"–	K 4856	–"–	1929	LMS 11642
130	K4	NBL 20071	1913	1932	LMS 10185
131	–"–	NBL 20072	–"–	1932	LMS 10186
1	D5	NBL 20073	–"–	1956	LMS 12494
2	–"–	NBL 20074	–"–	1932	LMS 12495
25	–"–	NBL 20075	–"–	1932	LMS 12496
26	–"–	NBL 20076	–"–	1935	LMS 12497
27	–"–	NBL 20865	1914	1932	LMS 12498
28	–"–	NBL 20866	–"–	1957	LMS 12499
132	K4	NBL 20867	–"–	1932	LMS 10187
133	–"–	NBL 20868	–"–	1932	LMS 10188
92	L4	K 5042	–"–	1934	LMS 11643
93	–"–	K 5043	–"–	1932	LMS 11644
38	M1	K 5119	1915	1930	LMS 1180
39	–"–	K 5120	–"–	1932	LMS 11081
51	G5	K 5121	–"–	1934	LMS 11559
52	–"–	K 5122	–"–	1930	LMS 11560
53	–"–	VF 3174	1916	1936	LMS 11561
54	–"–	VF 3175	–"–	1931	LMS 11562
40	M1	VF 3176	–"–	1930	LMS 11082
41	–"–	VF 3177	–"–	1932	LMS 11083
42	–"–	K 5172	–"–	1930	LMS 11084
43	–"–	K 5173	–"–	1931	LMS 11085
19	D5	K 5195	1918	1932	LMS 12500
20	–"–	K 5196	–"–	1957	LMS 12501
21	–"–	K 5197	–"–	1930	LMS 12502
22	–"–	K 5198	–"–	1930	LMS 12503
23	–"–	NBL 21993	–"–	1932	LMS 12504
24	–"–	NBL 21994	–"–	1932	LMS 12505
29	–"–	NBL 21995	–"–	1930	LMS 12506
30	–"–	NBL 21996	–"–	1935	LMS 12507
31	–"–	NBL 22572	1920	1950	LMS 12508
32	–"–	NBL 22573	–"–	1956	LMS 12509
33	–"–	NBL 22574	–"–	1957	LMS 12510
34	–"–	NBL 22575	1920	1932	LMS 12511
35	–"–	NBL 22576	–"–	1932	LMS 12512
115	N1	K 5292	–"–	1935	LMS 11100
116	–"–	K 5293	–"–	1935	LMS 11101
117	–"–	K 5294	1920	1934	LMS 11102
118	–"–	K 5295	–"–	1940	LMS 11103
119	–"–	K 5296	1921	1935	LMS 11104

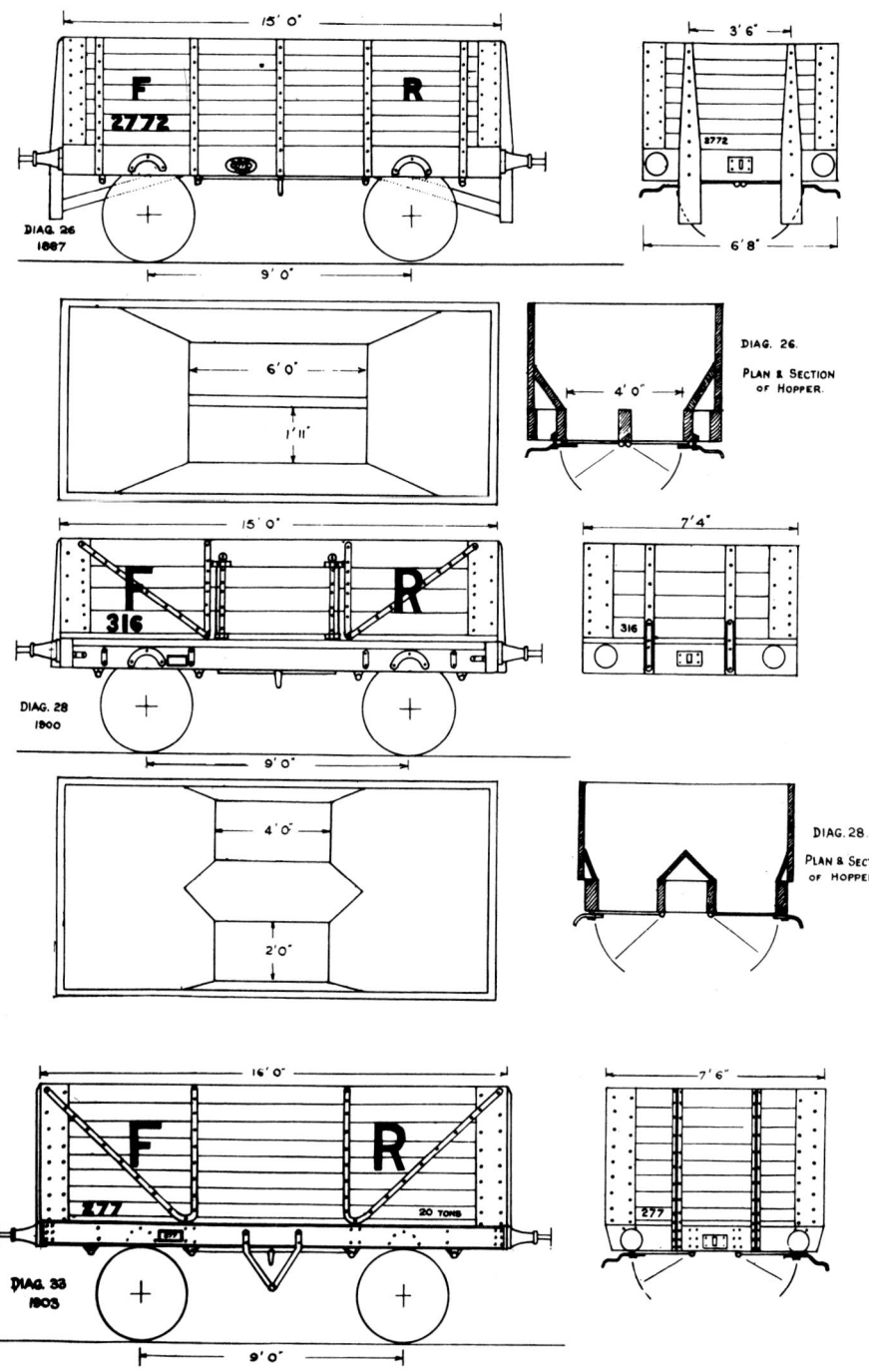

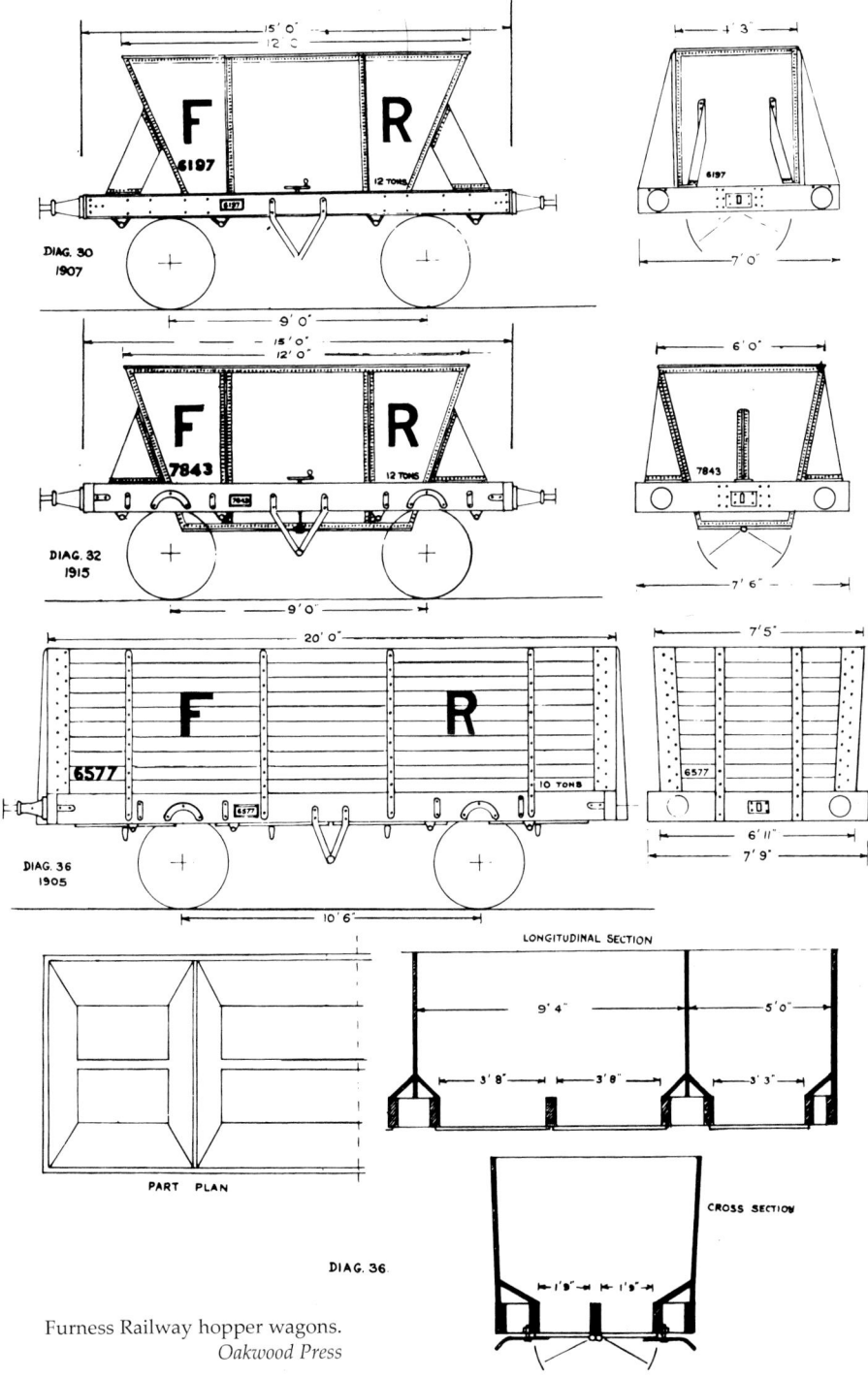

Furness Railway hopper wagons.
Oakwood Press

Appendix Two

Gradients
Main Line

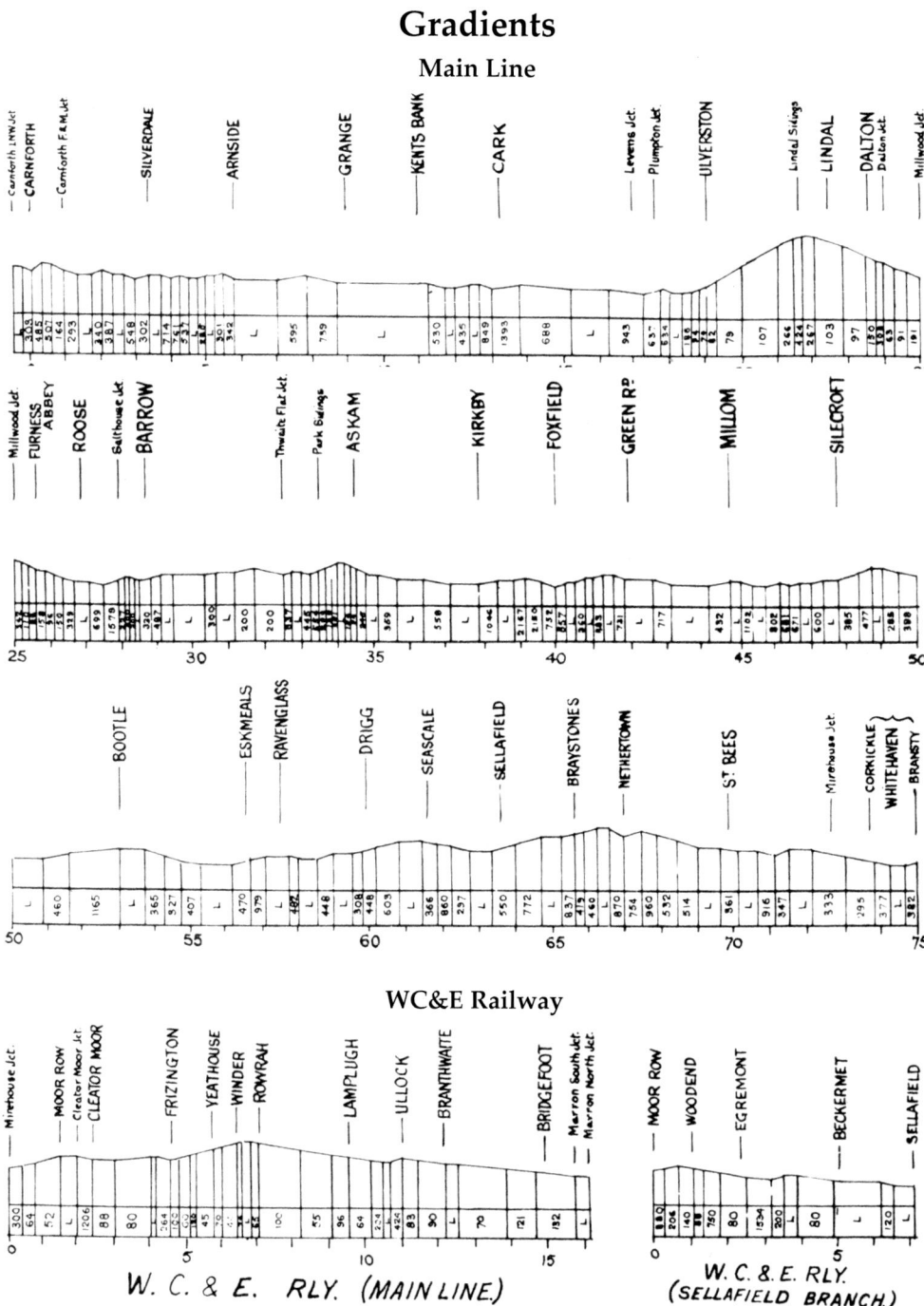

GRADIENTS

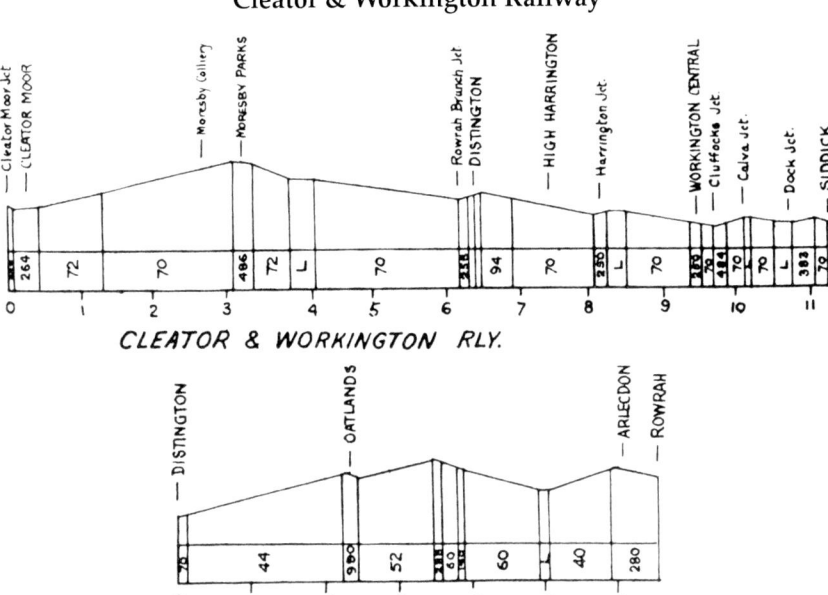

Index

Andrew Barclay, Sons & Co. 125, 129
Arkholm 44
Arnside 41, 172
Arnside to Hincaster line 42, 45, 48, 69
Askam 18, 19, 23, 63
Aslett, Alfred 68, 70, 74, 81, 131, 157

Baird's Line (*see* Rowrah & Kelton Fell Mineral Railway)
Baron Leconfield 89
Barrow 9, 17, 24, 43, 45, 63, 133, 137;
  Central station 63, 64, 99, 163;
  chosen as terminus of FR 11; early industry 17; mayors and aldermen 20, 28; original station 65
Barrow Docks 11-15, 17, 40, 45, 65, 67, 81, 102, 127, 144;, floating dock 74, 144;, opening 20
Barrow Haematite Steel Company 19, 20, 43, 46, 88, 105
Barrow Shipbuilding Company 12
Board of Trade 58, 66, 151, 157
Brakes on trains 66, 68, 85, 98, 103, 111, 115, 117, 118, 119, 120, 130, 137, 153, 155, 157, 158, 159, 165, 171; brake coaches 151, 153, 155, 157, 160
Bransty Tunnel 50, 52, 166; Sunday maintenance 83; telegraph 53; tolls 58
Brassey & Field 12

Bridgefoot 60
Broughton 31, 32, 37, 38, 50, 53, 165;
  W&FJR shed 54
Bury, Curtis & Kennedy 28, 99, 100

Cambrian Railways 68, 99, 112, 119, 129, 157
Carnforth 34, 37, 38, 63, 79, 117, 132, 133, 172; sheds 38, 87, 163
Carnforth to Wennington Line 39, 45, 87
Cleator & Workington Railway 61, 89-98, 140; main line opens 91; northern extension 91, 92
Cleator Moor 43, 57, 90, 115, 120, 123, 135, 137
Coal and coke traffic 25, 34, 42, 45, 48, 65, 68, 84, 94, 115, 161; Admiralty trains 79; rates 61
Cockermouth & Workington Railway 40, 49, 59, 89
Coniston Branch 32, 36, 37, 39, 84, 106, 108, 139, 141
Cook, Henry 42, 68
Copper mining 32, 36, 37, 39
Corkickle 52, 57, 58, 79, 86
Currey, Benjamin 26, 32

Dalton 24, 27, 28, 29, 31, 170; tunnel 33

Distington 61, 87, 91
Doubling line 31, 37, 39, 58, 59
Duddon estuary 21, 23, 40, 42, 50, 53
Duddon viaduct 43, 44, 45, 66, 74
Duke of Buccleuch, Walter Montagu Douglas Scott 23, 24, 25, 32, 42
Duke of Devonshire, saloon coach 151, 153; Spencer Compton Cavendish 66, 79; William Cavendish (*see also* Earl of Burlington) 14, 21, 46, 66

E. B. Wilson 108, 145
Earl of Burlington (*later* Duke of Devonshire) 21, 23, 24, 25, 27, 32
Earl of Lonsdale, William Lowther 24, 27, 40, 42, 49, 50, 53, 57, 89, 91
Early Railways/tramways 10, 24
Egremont 57, 58, 62, 85, 86
Eskmeals 69, 79

First World War 79-81, 94, 125
Fletcher, Jennings & Co. 95, 110, 114, 125, 128, 145
Foxfield 50, 51, 63
Franco-British Exhibition 72, 73
Furness Abbey 9, 11, 27; Hotel 28, 29
Furness Railway company; 1862 agreement with MR 39; early

extension plans 28; established 11; prospectus 25; established powers 39, 42, 48, 84; takeover of W&FJR 42, 53; works and offices 46, 47, 68, 141, 153, 162-163, 172; first aid classes 74; proposed Seascale to Egremont line 62;, rumour of MR takeover 65; to work C&WR 90

Gilgarron Branch 61, 88
Gradients 85, 86, 92, 93, 190-191
Caledonian, West Cumberland & Furness Railway 10, 21
Grange-over-Sands 21, 34, 119
Greenodd 31, 39, 43

Heysham 72
Hindpool Ironworks 18
Hodbarrow Mine 19, 41, 68

Irish Cattle traffic 162
Iron and Steel traffic 14, 23, 24, 25, 34, 37, 40, 42, 45, 46, 59, 66, 70, 84, 87, 94, 161; rates 31, 61
Iron Mining 18, 46/47, 68
James Little & Co. 46, 72

Kendal 39, 42, 69, 84, 120, 139, 140
Kent viaduct 34, 35, 74; reconstruction 80
Kitson & Co. 118, 127, 138, 139, 140

Lakeside Branch 39, 43, 84, 88, 108, 119, 139, 141
Lamplugh 59, 60, 61, 123
Lancashire & Yorkshire Railway 40, 100, 112, 134, 137, 147, 159
Lancaster 10, 32, 37, 53, 134, 140
Lancaster & Carlisle Railway 25, 27, 31, 32, 34
Lancaster & Preston Junction Railway 23
Lancaster Port Authority 11, 40; decline of 12
Leven Junction 39
Leven viaduct 31, 34, 42, 74, 169; reconstruction 80
Lindal 18, 31, 36, Bank 47, 85, 122, 123, 127; sidings 79, 112, 167
Livery, goods stock 162; locomotive 28; passenger stock 160; Ramsden saloon coach 155
London & North Western Railway 37, 40, 46, 59, 61, 62, 69, 79, 84, 90, 100, 121, 128, 134, 139, 145, 148, 158; running powers 40
London Midland Scottish Railway 81, 88, 97, 109, 112, 116, 118, 119, 123, 125, 127, 130, 131, 134, 135, 137, 138, 139, 158, 160
Lowca Coal Co. 87, 88, 92, 95
Lowca Engine Co. 97

Mail services 66, 79, 81, 82, 141, 159, 169, 172
Marron 59, 85, 86
Maryport 37, 90
Maryport & Carlisle Railway 23, 25, 37, 49, 50, 61, 79, 91
Midland Railway 13, 17, 39, 45, 69, 72, 74, 79, 87, 129, 158; running powers 39
Millom 19, 40, 41, 45, 50, 53, 74, 85
Milnthorpe 27, 32, 39, 40
Monkmoors halt 79
Moor Row 54, 57, 58, 85, 86; sheds 87, 95, 122, 124, 126, 127, 163
Morecambe 39, 48, 69, 83, 119, 139
Morecambe Bay, enclosure by embankments 21-24
Moss Bay Ironworks 90, 92, 97
Motor traffic 88

Nasmyth, Wilson & Co. 115, 137
Naval Construction & Armaments Co. 69
Newcastle-on-Tyne 72
North British Locomotive Co. 116, 118, 134, 137
North Eastern Railway 42, 72, 79, 159
North Lonsdale Ironworks 47

Parliamentary trains 151
Passenger traffic 34, 45/46, 50, 58, 65, 66, 68, 70, 83-87, 88, 93, 95, 115, 117, 119, 120, 127, 131, 132, 139, 157
Pettigrew, William Frank 70, 99, 104, 111, 114, 116, 117, 126, 132, 133, 135; retires 81
Preston & Wyre Railway 10, 22
Preston Street 49, 52, 165; engine shed 54

R. & W. Hawthorn 107, 114, 120, 126, 145, 147
Rail Motors 141-143, 159
Ramsden, Frederic James 74, 79
Ramsden, James 12, 14, 18, 20, 28, 32, 46, 66, 74; becomes director of FR 42; champions Barrow Docks 17; death 68; saloon coach 155
Ravenglass 32, 50, 51, 167
Roa Island (Piel Branch) 10, 24, 25, 30, 84;causeway 11, 34
Roadstead, Piel 9, 11, 24, 27, 31
Robert Stephenson & Co. 125
Rowrah 61, 86, 91
Rowrah & Kelton Fell Mineral Railway (Baird's Line) 88, 93
Rutherford, David L. 72, 74, 81, 140

Schneider & Hannay 17, 18
Schneider, William Henry 17, 20, 23, 43
Seascale 55, 62

Second class abolished 69
Sellafield 43, 59, 85
Seward, Mr A. B. 61
Sharp, Stewart & Co. 99, 104, 105, 106, 107, 109, 111, 115, 116, 119, 122, 127, 130, 132, 135
Silecroft 55
Silverdale 34, 73
Slate trade 23/24, 25
Smith, John Abel 10, 11, 25, 28, 31, 33
South Durham & Lancashire Union Railway 34
Spanish Iron Ore 14, 47, 68
Stage coaches 10, 32, 50, 53
Steamers on the lakes 16, 17, 53
Steamer services, Liverpool and Blackpool 71; Fleetwood 10, 11, 16, 23, 28, 31, 32, 53; Ireland and Isle of Man 13, 17, 39, 46, 72, 129
Stephenson, George 21, 49
Superheating 118, 130, 133, 134, 140

Tebay 34, 48, 65, 72, 84
Tourism 28, 39, 46, 47, 69, 72, 76-78; coach and steamer tours 69, 70, 72

Ulverston 9, 10, 21, 24, 25, 26, 27, 32, 33, 34, 35, 43, 47, 74, 105
Ulverston & Lancaster Railway 10, 33, 34, 37, 53; opens 34; purchased by Furness Railway 39
Ulverston Canal 10, 11, 31, 47

Vickers, Sons & Maxim (*also* Barrow Shipbuilding *and* Naval Construction & Armaments Co.) 12, 71, 79, 162
Vulcan Foundry 126, 127, 139, 147

W. Fairbairn 102, 103
Walker, John 23, 24
Walney Channel 12, 13, 24, 25, 65, 74
Whitehaven 21, 26, 37, 40, 49, 50, 58, 63, 69, 74, 79, 84, 116, 117, 132, 140, 172; harbour 57; harbour line 52, 54, 110
Whitehaven & Furness Junction Railway 26, 31, 32, 40, 49-55, 100, 107, 108, 109, 114, 120, 145-148, 151; first train to FR 50
Whitehaven Iron Company 18, 129
Whitehaven Junction Railway 40, 49
Whitehaven, Cleator & Egremont Railway 43, 48, 53, 54, 57-62, 95, 123, 125, 128, 135; becomes jointly owned by LNWR and FR 62
Windermere, Lake 39, 69
Wordsworth, William 27
Workington 49, 69, 79, 89, 90
Workmen's/Miner's services 79, 83, 87, 92, 94, 95, 139, 158, 159, 160